T0327328

Ultra Wideband Signals and Systems in Communication Engineering

Second Edition

Ultra Wideband Signals and Systems in Communication Engineering

Second Edition

M. Ghavami
King's College London, UK

L. B. Michael
Japan

R. Kohno
Yokohama National University, Japan

John Wiley & Sons, Ltd

Other Wiley Editorial Offices

John Wiley & Sons Inc., 111 River Street, Hoboken, NJ 07030, USA

Jossey-Bass, 989 Market Street, San Francisco, CA 94103-1741, USA

Wiley-VCH Verlag GmbH, Boschstr. 12, D-69469 Weinheim, Germany

John Wiley & Sons Australia Ltd, 42 McDougall Street, Milton, Queensland 4064, Australia

John Wiley & Sons (Asia) Pte Ltd, 2 Clementi Loop #02-01, Jin Xing Distripark, Singapore 129809

John Wiley & Sons Canada Ltd, 6045 Freemont Blvd, Mississauga, ONT, L5R 4J3, Canada

Wiley also publishes its books in a variety of electronic formats. Some content that appears in print
may not be available in electronic books.

British Library Cataloguing in Publication Data

A catalogue record for this book is available from the British Library

ISBN 978-0-470-02763-9 (HB)

Typeset by Sunrise Setting Ltd, Torquay, Devon, UK.

This book is printed on acid-free paper responsibly manufactured from sustainable forestry in which
at least two trees are planted for each one used for paper production.

Contents

Preface

In the two years since this book was first published, ultra wideband (UWB) has advanced and consolidated as a technology, and many more people are aware of the possibilities for this exciting technology. We too have expanded and consolidated materials in this second edition in the hope that 'Ultra Wideband: Signals and Systems in Communication Engineering' will continue to prove a useful tool for many students and engineers to come to an understanding of the basic technologies for UWB.

In this book we focus on the basic *signal processing* that underlies current and future UWB systems. By looking at signal processing in this way, we hope that this text will be useful even as UWB applications mature and change or regulations regarding UWB systems are modified. The current UWB field is extremely dynamic, with new techniques and ideas being presented at every communications and signal-processing conference. However, the basic signal-processing techniques presented in this text will not change for some time to come. This is because we have taken a somewhat theoretical approach, which we believe is longer lasting and more useful to the reader in the long term than an up-to-the-minute summary that is out of date as soon as it is published.

We restrict our discussion in general to *ultra wideband communication*, looking in particular at *consumer communication*. What we mean by this is that although there are many and varied specialized applications for UWB, particularly for the military, we assume that the majority of readers will either be in academia or in industry. In any case, as this is a basic text, aimed mostly at the upper undergraduate or graduate student, these basics should stand the reader in good stead to be able to easily understand more advanced papers and make a contribution in this field for themselves.

We are painfully aware of the depth and breadth of this field, and regretfully pass on interesting topics such as UWB radar, including ground penetrating radar, and most military applications. For the former there is already a great deal of information available, while for the latter most material is classified.

The introduction to this book presents a brief look at why UWB is considered to be an exciting wireless technology for the near future. We examine Shannon's famous capacity equation and see that the large bandwidth promises possibilities for high-data-rate communication. A quick overview of the regulatory situation is presented.

Chapter 1 presents the basic properties of UWB. We examine the power spectral density, basic pulse shape, and spectral shape of these pulses. The regulatory requirements laid down by the Federal Communications Commission are briefly described. Why UWB is considered to be a multipath resistant form is also examined, and such basic figures of merit such as capacity and speed of data transmission are considered. We finish the chapter with a look at the cost, size, and power consumption that is forecast for UWB devices and chipsets.

Chapter 2 examines in detail how to generate basic pulse waveforms for UWB systems, for the simple Gaussian pulse shape. An introduction to damped sine waves and the difference between them and Gaussian waveforms is presented. Armed with this information, the reader can now proceed to more complex waveforms and theory associated with UWB signals and systems. We examine how to design pulses to fit spectral masks, such as mandated by regulators, or to avoid interference to other frequency bands.

Chapter 3 looks at different signal-processing techniques for UWB systems. The chapter begins with a review of basic signal-processing techniques, including both time and frequency domain techniques. The Laplace, Fourier, and z-transforms are reviewed and their application to UWB is discussed. Finally, some practical issues, such as pulse detection and amplification, are discussed.

The wireless indoor channel, and how it should be modeled for UWB communications, is considered in Chapter 4. Following our basic pattern we define and explore basic concepts of wideband channel modeling, and show a simplified UWB multipath channel model which is amendable to both theoretical analysis and simulation. Path loss effects and a two-ray model are presented. A frequency domain autoregressive model is discussed and, finally, IEEE proposals for a UWB channel model are explained.

Chapter 5 takes a look at some of the fundamental communication concepts and how they should be applied to UWB. First, modulation methods applicable to UWB are presented. A basic communication system consisting of transmitter, receiver, and channel is discussed. Since most consumer communication systems do not consist of only one user, multiple access techniques are introduced. The simple capacity of a UWB system is derived. Since other wireless consumer communication systems have already become popular, a comparison between UWB and other wideband techniques is included. Finally, the chapter ends with a look at interference to and from UWB systems.

In Chapter 6, which is in many ways an extension of Chapter 2 but requiring many of the concepts presented in Chapters 3 to 5, more complex pulse shapes and their use in a communication system are explained. An extensive treatment of the more complex orthogonal pulses, including Hermite pulses, prolate spheroidal wave functions, and wavelet packets is presented.

Chapter 7 is concerned with UWB antennas and arrays of antennas. This is considered one of the most difficult problems that must be overcome before the widespread commercialization of UWB devices takes place. Antenna fundamentals are first introduced, including Maxwell's equations for free space, antenna field regions, directivity, and gain. The suitability of conventional antennas for UWB transmission and reception is discussed in detail. More suitable impulse antennas are then introduced. Arrays of antennas and beamforming for UWB systems are given a brief treatment.

Positioning and location, using both traditional techniques and UWB, are discussed in Chapter 8. Traditional location systems are first introduced and their pros and cons discussed. The advantages of UWB, particularly the extremely precise positioning that is theoretically possible, are examined. Finally, several possible scenarios are discussed where the precise location capabilities and high data rate of UWB can be combined to produce some new and exciting applications.

New applications made possible by UWB technology are among the most exciting reasons to use UWB. Chapter 9 has a brief look at some applications that use UWB technology, as well as an overview of some chipsets and possible future UWB products. Emphasis is on consumer communication and medicine; however, military applications are also given a brief treatment.

Chapter 10, an additional chapter for the second edition, presents an introduction and overview of the main UWB standards bodies. In particular, the IEEE 802.15.3a and IEEE 802.15.4a efforts are summarized. The two main physical layer proposals for UWB, direct sequence UWB and multiband UWB, and their respective advantages are then presented in detail.

Chapter 11 presents advanced topics in UWB communication systems, and is also an addition for the second edition. This chapter looks at novel communication systems that have matured recently. In particular, UWB ad-hoc and sensor networks, UWB vehicular radars and the effects of interference with Wi-Max are examined.

For the reader who wants a fast-track understanding of UWB and some knowledge of the current situation, we recommend the introduction, Chapter 1 (Basic properties of UWB signals and systems), Chapter 9 (Applications), and Chapter 10 (UWB communication standards).

For students who want to look at UWB in more detail, they should then proceed to look at Chapter 2 (Generation of UWB waveforms), Chapter 3 (Signal processing techniques for UWB systems), and then Chapters 4 through to 8 as required. We have strived to make each chapter complete in itself as far as possible and provide as much basic theory as practicable, including derivations where appropriate. We have made constant reference to the literature, a significant part of which is covered here.

As an extra resource we have set up a companion website for our book containing a solutions manual, Matlab programs for the examples and problems, and a sample chapter. Also, for those wishing to use this material for lecturing purposes, electronic

versions of most of the figures from our book are available. Please take a look at
http://www.wiley.com/go/ghavami.

We hope that you will find this book useful as both a reference, a learning tool,
and a stepping stone to further your own efforts in this exciting field.

M. Ghavami
L. B. Michael
R. Kohno

London, January 2007

Acknowledgments

The authors would like to thank the following people for their efforts and contributions to the second edition of *Ultra Wideband Signals and Systems in Communication Engineering*:

– Sarah Hinton, our editor, for her tireless and unending efforts to make this publication timely and well received, as well as for helping us with the ins and outs of writing a textbook;

– Dr X. Chu, Dr F. Heliot, S. Ciolino, K. Sarfaraz, Dr R. S. Dilmaghani, W. Horie, N. Riaz and K. Kang (King's College London) for their valuable contributions.

M. Ghavami would like to thank:

my wife Mahnaz and my children Navid and Nooshin who have suffered the long period of preparation of this book and who have been continually supportive.

L. B. Michael would like to thank:

my wife and children for their support and patience during the weekends and nights while I was preparing and editing material for this book.

List of Figures

List of Tables

Introduction

In this chapter we present a general background to UWB and try to explain, without resorting to too many equations, the reasons why UWB is considered to be an exciting and breakthrough technology. We place UWB in its historical context and show that, while UWB is not necessarily entirely new in either the concept or the signal-processing techniques used, given the recent emphasis on wireless communication on sinusoidal systems, UWB does present a paradigm shift for many engineers.

We believe the current (and for the foreseeable future) emphasis on low power, low interference and low regulation makes the use of UWB an attractive option for current and future wireless applications.

I.1 ULTRA WIDEBAND OVERVIEW

Historically, UWB radar systems were developed mainly as a military tool because they could 'see through' trees and beneath ground surfaces. However, recently, UWB technology has been focused on consumer electronics and communications. Ideal targets for UWB systems are low power, low cost, high data rates, precise positioning capability, and extremely low interference.

Although UWB systems are years away from being ubiquitous, the technology is changing the wireless industry today. UWB technology is different from conventional narrowband wireless transmission technology – instead of broadcasting on separate frequencies, UWB spreads signals across a very wide range of frequencies. The typical sinusoidal radio wave is replaced by trains of pulses at hundreds of millions of pulses

Ultra Wideband Signals and Systems in Communication Engineering Second Edition
M. Ghavami, L. B. Michael and R. Kohno © 2007 John Wiley & Sons, Ltd

per second. The wide bandwidth and very low power make UWB transmissions appear as background noise.

I.2   A NOTE ON TERMINOLOGY

The name ultra wideband is an extremely general term to describe a particular technology. Many people feel other names, such as pulse communications, may be more descriptive and suitable. However, UWB has become the term by which most people refer to ultra wideband technology.

The question then arises as to how to spell UWB. Is it 'ultrawideband', 'ultra-wideband', 'ultra wide band', 'ultrawide band', or 'ultra wideband'? In this text, quite arbitrarily, we decide to use the term *ultra wideband*. Our reasoning is that the term wideband communication has become very common in recent years and is one that most people are familiar with. To show that UWB uses an even larger bandwidth the extra large 'ultra' is prefixed; however, both 'ultrawideband' and 'ultra-wideband' seem unwieldy, so we use ultra wideband. Many people may disagree about our choice, even vehemently. We accept their arguments and suggest that time will show the most popular choice for UWB.

I.3   HISTORICAL DEVELOPMENT OF UWB

Most people would see UWB as a 'new' technology, in the sense that it provides the means to do what has not been possible before, be that the use of high data rates, smaller, lower powered devices or, indeed, some other new application. However, UWB is, rather, a *new engineering technology* in that no new physical properties have been discovered.

However, the dominant method of wireless communication today is based on sinusoidal waves. Sinusoidal electromagnetic waves have become so universal in radio communications that many people are not aware that the first communication systems were in fact pulse-based. It is this paradigm shift for today's engineers from sinusoids to pulses, that requires the most shift in focus.

In 1893 Heinrich Hertz used a spark discharge to produce electromagnetic waves for his experiment. These waves would be called colored noise today. Spark gaps and arc discharges between carbon electrodes were the dominant wave generators for about 20 years after Hertz's first experiments.

However, the dominant form of wireless communications became sinusoidal, and it was not until the 1960s that work began again in earnest for time domain electromagnetics. The development of the sampling oscilloscope in the early 1960s and the corresponding techniques for generating sub-nanosecond baseband pulses sped up the development of UWB. Impulse measurement techniques were used to characterize the transient behavior of certain microwave networks.

From measurement techniques the main focus moved to develop radar and communications devices. In particular, radar was given much attention because of the accurate results that could be obtained. The low-frequency components were useful in penetrating objects, and *ground-penetrating radar* was developed. See [1] and [2] for more details about UWB radar systems.

In 1973 the first US patent was awarded for UWB communications [3]. The field of UWB had moved in a new direction. Other applications, such as automobile collision avoidance, positioning systems, liquid-level sensing, and altimetry, were developed. Most of the applications and development occurred in the military or in work funded by the US Government under classified programs. For the military, accurate radar and low probability of intercept communications were the driving forces behind research and development.

It is interesting to note that in these early days UWB was referred to as *baseband*, *carrier-free* and *impulse* technology. The US Department of Defense is believed to be the first to have started to use the term *ultra wideband*.

The late 1990s saw the move to commercialize UWB communication devices and systems. Companies such as Time Domain [4] and in particular startups like XtremeSpectrum [5] were formed around the idea of consumer communication using UWB.

For further historical reading, the interested reader is referred to [6] and [7].

I.4 UWB REGULATION OVERVIEW

For the regulation of UWB around the world there are many organizations and government entities that set rules and recommendations for UWB usage. The structure of international radio-communication regulatory bodies can be grouped into international, regional, and national levels.

The International Telecommunication Union (ITU) is an impartial, international body where governments and the private sector work together on issues pertinent to telecommunication networks. The group that undertakes the work on UWB was created in 2002 to study the compatibility between UWB and other communication services. Most of the telecommunication services that occupy the allocated spectrum would prefer to keep UWB out of their frequency range.

At the regional level, the Asia-Pacific Telecommunity (APT) is an international body that sets recommendations and guidelines of telecommunications in the Asia-Pacific region. The European Conference of Postal & Telecommunications Administrations (CEPT) has created a task group under the Electronic Communications Committee (ECC) to draft a proposal regarding the use of UWB for Europe.

At the national level, the USA was the first country to legalize UWB for commercial use. The rules are meant to protect existing radio devices, particularly those classified as safety devices, such as aviation systems and the global positioning system (GPS). In March 2005, the Federal Communications Commission (FCC) granted a waiver that will lift certain limits on UWB.

In the UK, the regulatory body, called the Office of Communications (Ofcom), opened a consultation on UWB matters in January 2005. The consultation consisted of 15 questions, asking opinions from those who are affected by the UWB technology.

Ofcom sees UWB as a positive technology that if correctly regulated can bring economic growth to the UK.

The regulatory body that set the policy on UWB in Japan is called the Ministry of Internal Affairs and Communications (MIC). The first interim report was published in March 2004, which drafted two proposals on the limit of UWB emission and addressed the issue of interference. One proposal set the limit of UWB to be the same as that set by the FCC, and the other one set a slightly more restrictive limit. Overall, the report suggests that indoor use is the best environment for UWB application.

Looking through each country's regulatory status gives the impression that all countries monitor the UWB development in the USA closely while they themselves have yet to make any significant progress to catch up to the level of UWB regulation on UWB. There can be two reasons for this. First, UWB has already been developed in the USA under the classification of 'military use' for many years. The second reason can be classified as politics. Many current industry wireless manufacturers and wireless service providers voice their concerns on the potential damage to their business loudly, leading the regulatory bodies to be conservative in their outlook.

I.4.1   Basic definitions and rules

The FCC rules provide the following definitions for UWB signaling:

- *UWB bandwidth:* UWB bandwidth is the frequency band bounded by the points that are 10 dB below the highest radiated emission, as based on the complete transmission system including the antenna. The upper boundary is designated f_h and the lower boundary is designated f_l. The frequency at which the highest radiated emission occurs is designated f_m.

- *Center frequency:* The center frequency f_c is the average of f_l and f_h, that is,

$$f_c = \frac{f_l + f_h}{2} \tag{I.1}$$

- *Fractional bandwidth* (FB): The fractional bandwidth is defined as

$$\mathrm{FB} = 2\frac{f_h - f_l}{f_h + f_l} \tag{I.2}$$

- *UWB transmitter:* A UWB transmitter is an intentional radiator that, at any point in time, has a fractional bandwidth equal to or greater than 0.20 or has a UWB bandwidth equal to or greater than 500 MHz, regardless of the fractional bandwidth.

- *Equivalent isotropically radiated power* (EIRP): EIRP is the product of the power supplied to the antenna and the antenna gain in a given direction relative to an isotropic antenna. EIRP refers to the highest signal strength measured in any direction and at any frequency from the UWB device.

The first set of FCC key regulations for all UWB systems are as follows:

- No toys, and no operation on an aircraft, ship or satellite.

- Emissions from supporting digital circuitry is considered separately from the UWB portion, and is subject to existing regulations, not new UWB rules.

- The frequency of the highest emission, f_m, must be within the UWB bandwidth.

- Other emissions standards apply as cross-referenced in the UWB rules, such as conducted emissions into AC power lines.

- Emissions below 960 MHz are limited to the levels required for unintentional radiators.

- Within a 50 MHz bandwidth centered on f_m, peak emissions are limited to 0 dBm EIRP.

- UWB radar, imaging and medical system operation must be coordinated. Dates and areas of operation must be reported, except in the case of emergency. These systems must also have a manual switch (local or remote) to turn the equipment off within 10 s of actuation.

Discussion continues on UWB measurement methodology and these first rules are likely to change.

I.5 KEY BENEFITS OF UWB

The key benefits of UWB can be summarized as:

(1) high data rates;

(2) low equipment cost;

(3) multipath immunity;

(4) ranging and communication at the same time.

We will expand on these benefits in the coming chapters, but first we give a brief overview.

The high data rates are perhaps the most compelling aspect from a user's point of view and also from a commercial manufacturer's position. Higher data rates can enable new applications and devices that would not have been possible up until now. Speeds of over 100 Mbps have been demonstrated, and the potential for higher speeds over short distances is there. The extremely large bandwidth occupied by UWB gives this potential, as we show in the next section.

The ability to directly modulate a pulse onto an antenna is perhaps as simple a transmitter as can be made, leading many manufacturers to get excited by the possibilities for extremely cheap transceivers. This is possible by eliminating many of the components required for conventional sinusoidal transmitters and receivers.

The narrow pulses used by UWB, which also give the extremely wide bandwidth, if separated out provide a fine resolution of reflected pulses at the receiver. This is important in any wireless communication, as pulses (or sinusoids) interfering with each other are the major obstacle to error-free communication.

Finally, the use of both precise ranging (object location) and high-speed data communication in the same wireless device presents intriguing possibilities for new devices and applications. Simultaneous automotive collision avoidance radar and communication giving accident-free smooth traffic flow, or games where the players' position can be precisely known and a high-speed wireless link seamlessly transfers a video signal to the players' goggles may seem the stuff of science fiction, but with UWB the possibilities for these and other applications are there, right now.

I.6 UWB AND SHANNON'S THEORY

Perhaps the benefits and possibilities of UWB can be best summarized by examining Shannon's famous capacity equation. This equation will be familiar to anyone who has studied communication or information theory. Capacity is important, as more demanding audio-visual applications require higher and higher bit rates.

Shannon's equation is expressed as

$$C = B \log \left(1 + \frac{S}{N} \right) \tag{I.3}$$

where C is the maximum channel capacity, with units bits per second, B is the channel bandwidth in hertz, S is the signal power in watts, and N is the noise power also in watts.

This equation tells us that there are three things that we can do to improve the capacity of the channel. We can increase the bandwidth, increase the signal power or decrease the noise. The ratio S/N is more commonly known as the *signal-to-noise ratio* (SNR) of the channel. We also can see that the capacity of a channel grows linearly with increasing bandwidth B, but only logarithmically with signal power S.

The UWB channel has an abundance of bandwidth and in fact can trade off some of the bandwidth for reduced signal power and interference from other sources. Thus, from Shannon's equation we can see that UWB systems have a great potential for high-capacity wireless communications.

Another way of looking at wireless communication is the tradeoffs between:

- the distance between transmitter and receiver;

- simultaneous communication for many users;

- sending the data very quickly;

- sending and receiving a large amount of data.

The first wireless communication systems, such as wireless communication at sea, were meant to communicate between ships separated by large distances. However, the amount of data that could be effectively transferred was extremely small and communication took a long time. Only one person can 'talk' using Morse code at a time. More recently, cellular telephone systems have simultaneous communication for many users as their forté. The distance between the base station and the user is limited to at most a few kilometers. It can be classified as a system where a moderate amount of data can be sent reasonably quickly. An UWB system is focused on the latter two attributes: a large amount of data that can be transmitted very quickly. This is at the expense of, in the main, distance. The precise tradeoffs are of course more complex and will depend upon the particular application.

I.7 CHALLENGES FOR UWB

While UWB has many reasons to make it an exciting and useful technology for future wireless communications and many other applications, it also has some challenges that must be overcome for it to become a popular and ubiquitous technology.

Perhaps the most obvious one to date has been regulatory problems. Wireless communications have always been regulated to avoid interference between different users of the spectrum. Since UWB occupies such a wide bandwidth, there are many users whose spectrum will be affected and they need to be convinced that UWB will not cause undue interference to their existing services. In many cases these users have paid to have exclusive use of the spectrum.

Other challenges include the industry coming to agreed standards for inter-operability of UWB devices. There are currently two camps of UWB supporters each with their own standard of UWB design, and currently there is no compromising ground to resolve this issue. This standard battle is a concern because in the near future consumers will be hesitant to choose which standard to buy and thus limit the potential of UWB market growth.

Many technical and implementation issues remain. The promise of low-cost devices is there, but the added complexity to combat interference and low-power operation may bring cost increases similar to current wireless devices.

I.8 SUMMARY

In this chapter we presented a general background to UWB and explained the reasons UWB is considered to be an exciting and breakthrough technology, particularly from the viewpoint of Shannon's famous capacity equation. We placed UWB in its historical background and showed the development of UWB from radar to communications applications. We briefly explained different international regulatory bodies and presented the basic definitions and rules of UWB signal systems. We showed the differences in the concept of the signal-processing techniques used for sinusoidal narrowband systems and those for pulse-based UWB systems.

Problems

Problem 1. Investigate the current regulations for UWB in your country. List other uses of the same wireless bandwidth.

Problem 2. Read and summarize a UWB journal or conference paper published before 1990. Discuss how the UWB technology described in that paper has changed. You may want to compare and contrast a more recent paper discussing the same topic.

Problem 3. Many wireless technologies, including UWB, were first used and developed by and for the military. Discuss your views on this progression of technology from the military to consumer markets. What are the possible pros and cons?

1

Basic properties of UWB signals and systems

1.1   INTRODUCTION

In this chapter the basic properties of UWB signals and systems are outlined, with details of each of the characteristics being explained in later chapters.

First, we examine the basic shape of an UWB pulse in the time domain and see what the spectrum content is of these pulses. Generally speaking, the extremely short pulses with fast rise and fall times have a very broad spectrum and a very small energy content. We examine the regulatory aspects of power output and frequency by spectral masks.

Next, we see that because UWB pulses are extremely short they can be filtered or ignored. They can readily be distinguished from unwanted multipath reflections because of the fine time resolution. This leads to the characteristic of multipath immunity.

Furthermore, the low-frequency components of the UWB pulses enable the signals to propagate effectively through materials such as bricks and cement.

The large bandwidth of UWB systems means extremely high data rates can be achieved, and we show that UWB systems have a potentially high spectral capacity.

UWB transmitters and receivers do not require expensive and large components such as modulators, demodulators, and intermediate frequency (IF) stages. This fact can reduce cost, size, weight, and power consumption of UWB systems compared with conventional narrowband communication systems.

Ultra Wideband Signals and Systems in Communication Engineering   Second Edition
M. Ghavami, L. B. Michael and R. Kohno   © 2007 John Wiley & Sons, Ltd

1.2 POWER SPECTRAL DENSITY

The power spectral density (PSD) of UWB systems is generally considered to be extremely low, especially for communication applications. The PSD is defined as

$$\text{PSD} = \frac{P}{B} \tag{1.1}$$

where P is the power transmitted in watts, B is the bandwidth of the signal in hertz, and the unit of PSD is watts/hertz.

Historically, wireless communications have only used a narrow bandwidth and can hence have a relatively high power spectral density. We can put this another way: since we know that frequency and time are inversely proportional, sinusoidal systems have narrow B and long time duration t. For a UWB system, the pulses have a short t and very wide bandwidth B. It is helpful to review some traditional wireless broadcast and communication applications, and to calculate their PSDs as shown in Table 1.1.

Table 1.1 PSD of some common wireless broadcast and communication systems.

System	Transmission power	Bandwidth	PSD [W/MHz]	Classification
Radio	50 kW	75 kHz	666 600	Narrowband
Television	100 kW	6 MHz	16 700	Narrowband
2G Cellular	500 mW	8.33 kHz	60	Narrowband
802.11a	1 W	20 MHz	0.05	Wideband
UWB	0.5 mW	7.5 GHz	6.670×10^{-8}	Ultra wideband

The energy used to transmit a wireless signal is not infinite and, in general, should be as low as possible, especially for today's consumer electronic devices. If we have a fixed amount of energy we can either transmit a great deal of energy density over a small bandwidth or a very small amount of energy density over a large bandwidth. This comparison is shown for the PSD of two systems in Figure 1.1. The total amount of power can be calculated as the area under a frequency–PSD graph.

For UWB systems, the energy is spread out over a very large bandwidth (hence the name ultra wideband) and, in general, is of a very low PSD. The major exception to this general rule of thumb is UWB radar systems, which transmit at high power over a large bandwidth. However, here we will restrict ourselves to the communications area.

One of the benefits of low PSD is a *low probability of detection*, which is of particular interest for military applications, such as covert communications and radar. This is also a concern for wireless consumer applications, where the security of data for corporations and individuals using current wireless systems is considered to be insufficient [8].

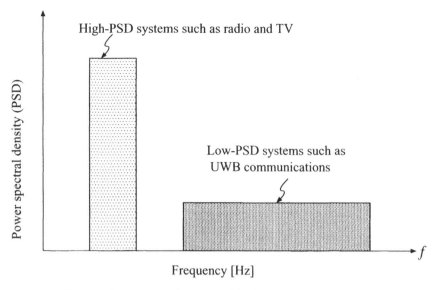

Fig. 1.1 Low-energy density and high-energy density systems.

1.3 PULSE SHAPE

A typical received UWB pulse shape, sometimes known as a *Gaussian doublet*, is shown in Figure 1.2. More details regarding Gaussian and other waveforms are discussed in Chapters 2 and 6. This pulse is often used in UWB systems because its shape is easily generated. It is simply a square pulse which has been shaped by the limited rise and fall times of the pulse and the filtering effects of the transmit and receive antennas. A square pulse can be easily generated by switching a transistor on and off quickly.

We show a simple pulse generator model in Figures 1.3 and 1.4, which demonstrate the creation of Gaussian doublets at a transmitter, antenna effects and reception. We start with a rectangular pulse in Figure 1.4(a). UWB pulses are typically of nanosecond or picosecond order. The fast switching on and off leads to a pulse shape which is not rectangular, but has the edges smoothed off. The pulse shape approximates the Gaussian function curve. A Gaussian function $G(x)$ is one which fits the well-known equation

$$G(x) = \frac{1}{\sqrt{2\pi\sigma^2}}e^{-x^2/2\sigma^2} \tag{1.2}$$

where Equation (1.2) is assumed to be zero-mean. This is the origin of the name Gaussian pulse, monocycle or doublet. A simple circuit for creation of the Gaussian doublet is shown in Figure 1.3. Transmitting the pulses directly to the antennas results in the pulses being filtered due to the properties of the antennas. This filtering operation can be modeled as a derivative operation [9]. The same effect occurs at

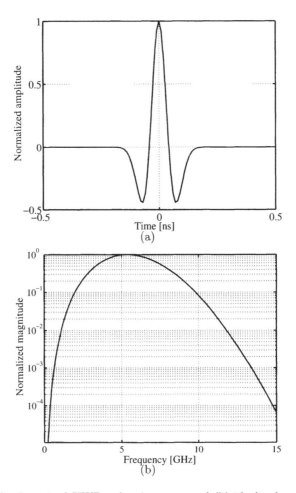

Fig. 1.2 (a) Idealized received UWB pulse shape p_{rx} and (b) idealized spectrum of a single received UWB pulse.

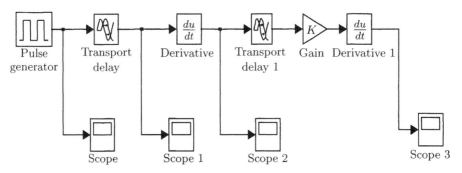

Fig. 1.3 A simple Matlab circuit model to create the Gaussian doublet.

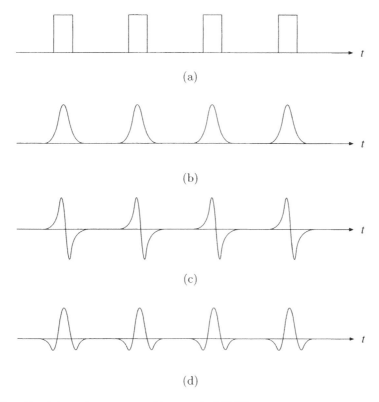

Fig. 1.4 Details of the pulses generated in a typical UWB communication system: (a) square pulse train; (b) Gaussian-like pulses; (c) first-derivative pulses; (d) received Gaussian doublets.

the receive antenna. Here, we model the channel as a delay and assume the pulse is amplified at the receiver.

In this chapter we will limit our discussion to this typical received pulse shape, which is assumed for the large majority of UWB research and is reasonably close to received pulses that have been measured. Details of pulse generation and a detailed discussion of pulse shaping can be found in Chapters 2 and 6.

This idealized received pulse shape p_{rx} can be written as [10]

$$p_{rx} = \left[1 - 4\pi \left(\frac{t}{\tau_m}\right)^2\right] e^{-2\pi(t/\tau_m)^2} \tag{1.3}$$

which is the equation used to generate the pulse shown in Figure 1.2(a). Here, τ_m is assumed to be 0.15. It should be mentioned that τ_m is the single parameter of Equation (1.3) and determines the time and frequency characteristics of the Gaussian doublet uniquely.

The spectrum of the Gaussian doublet is shown in Figure 1.2(b). The center frequency can be seen to be approximately 5 GHz, with the 3 dB bandwidth extending over several GHz. In comparison with narrowband or even wideband communication systems the large bandwidth is evident and, hence, the name ultra wideband communication can easily be inferred.

1.4 PULSE TRAINS

One pulse by itself does not communicate a lot of information. Information or data needs to be modulated onto a sequence of pulses called a *pulse train*.

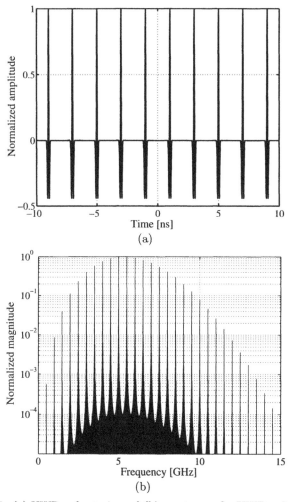

Fig. 1.5 (a) UWB pulse train and (b) spectrum of a UWB pulse train.

When pulses are sent at regular intervals, which is sometimes called the *pulse repetition rate* or the *duty cycle*, the resulting spectrum will contain peaks of power at certain frequencies. These frequencies are the inverse of the pulse repetition rate. These peak power lines are called *comb lines* because they look like a comb. See Figure 1.5(b) for an example.

These peaks limit the total transmit power undesirably. One method of making the spectrum more noise-like is to 'dither' the signal by adding a small random offset to each pulse, either delaying the pulse or transmitting slightly before the regular pulse time. The resultant spectrum from such a random offset is shown in Figure 1.6 and should be compared with Figure 1.5(b).

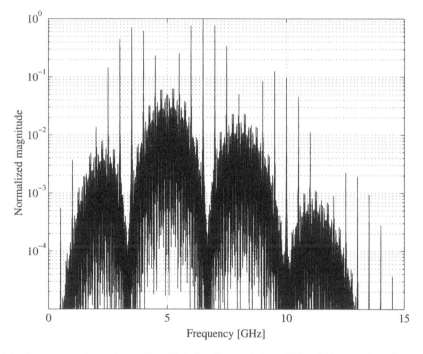

Fig. 1.6 Spectrum of a pulse train which has been 'dithered' by shifting pulses forward and backward of the original position.

As we will see in Chapter 5, by making this delay not completely random but cyclic according to a known pseudo-noise (PN) code, information can be modulated onto the pulse waveform. This is known as *pulse position modulation* (PPM) and has been investigated in different communication systems such as optical wireless communications.

1.5 SPECTRAL MASKS

The spectrum of a UWB signal is one of the major issues confronting the industry and governments for the commercial use of UWB. In fact, the very name ultra wideband suggests that the issue of spectrum is at the very heart of the UWB technology.

All radio communication is subject to different laws and regulations about power output in certain frequency bands. This is to prevent interference to other users in nearby or the same frequency bands.

UWB systems cover a large spectrum and interfere with existing users. In order to keep this interference to a minimum, the FCC and other regulatory groups specify *spectral masks* for different applications, which show the allowed power output for specific frequencies.

In Figure 1.7 an example is shown of the FCC spectral mask for indoor UWB systems. A large contiguous bandwidth of 7.5 GHz is available between 3.1 GHz and 10.6 GHz at a maximum power output of −41.3 dBm/MHz.

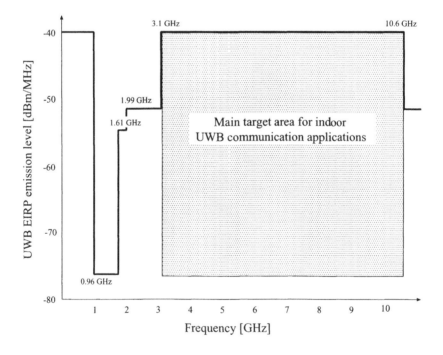

Fig. 1.7 Spectral mask mandated by FCC 15.517(b,c) for indoor UWB systems.

The major reasons for the extremely low allowed power output in the frequency bands 0.96–1.61 GHz is due to pressure from groups representing existing services, such as mobile telephony, GPS, and military usage. The allowed level of −41.3 dBm/ MHz itself is considered conservative, and many groups have lobbied for higher allowed power output.

1.6 MULTIPATH

In this section we will look at the effects of multipath, particularly in an indoor wireless channel, on the basic UWB pulse we have described. We will see that, because of the extremely short pulse widths, if these pulses can be resolved in the time domain then the effects of multipath, such as *inter-symbol interference* (ISI), can be mitigated.

Multipath is the name given to the phenomenon at the receiver whereby after transmission an electromagnetic signal travels by various paths to the receiver. See Figure 1.8 for an example of multipath propagation within a room. This effect is caused by reflection, absorption, diffraction, and scattering of the electromagnetic energy by objects in between the transmitter and the receiver. If there were no objects to absorb or reflect the energy, this effect would not take place and the energy would propagate outward from the transmitter, dependent only on the transmit antenna characteristics. However, in the real world, objects between the transmitter and the receiver cause the physical effects of reflection, absorption, diffraction, and scattering, and this gives rise to multiple paths. Due to the different lengths of the paths, pulses will arrive at the receiver at different times, with the delay proportional to the path length.

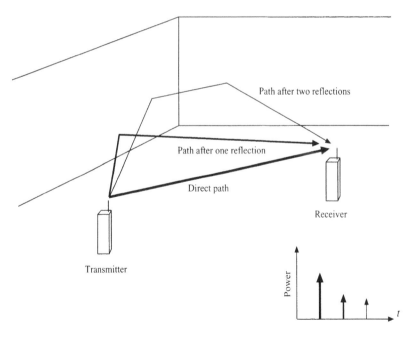

Fig. 1.8 A typical indoor scenario in which the transmitted pulse is reflected off objects within the room, thus creating multiple copies of the pulse at the receiver, with different delays.

UWB systems are often characterized as *multipath immune* or *multipath resistant*. Examining the pulses described previously, we can see that if pulses arrive within

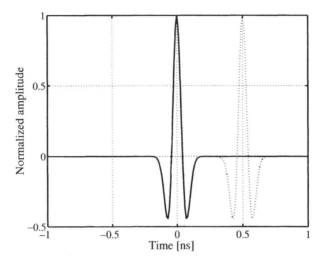

Fig. 1.9 Two pulses arriving with a separation greater than the pulse width will not overlap and will not cause interference.

one pulse width they will interfere, while if they are separated by at least one pulse width they will not interfere. If pulses do not overlap, then they can be filtered out in the time domain or, in other words, ignored. Assuming one symbol per pulse, they will not produce interference with the same symbol. Alternatively, the energy can be summed together by a *rake receiver*. Figures 1.9 and 1.10 demonstrate nonoverlapping and overlapping pulses, respectively.

Example 1.1
Assuming a received pulse shape similar to Figure 1.2, how much extra distance must a second pulse travel to not interfere with the original pulse? If the pulse width was halved, what would be the separation between multiple paths needed?

Solution
From Figure 1.2 the pulse width is approximately 0.4 ns. Using the relation that distance is the product of velocity with time travelled, $d = v \cdot t$, and since electromagnetic energy travels at a velocity of approximately 3×10^8 m/s, the extra distance travelled via the second path to avoid interference at the receiver is 12 cm. If the pulse width was halved, the required distance between multipaths to avoid interference would be halved also, to 6 cm.

As we can see from Example 1.1 the separation distance required between multipaths decreases with decreasing pulse width. This is one reason for smaller pulse widths, particularly in indoor environments.

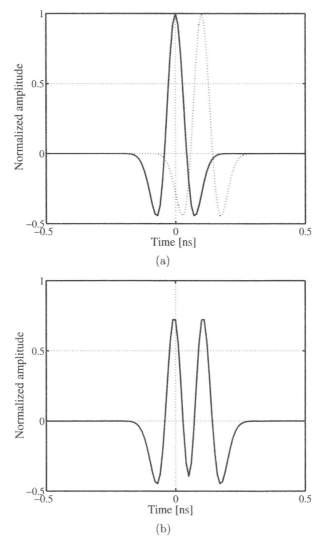

Fig. 1.10 (a) Two overlapping UWB pulses, and (b) the received waveform consisting of the overlapped pulses.

Another method to avoid multipath interference is to lower the duty cycle of the system. By transmitting pulses with time delays greater than the maximum expected multipath delay, unwanted reflections can be avoided at the receiver. This is inherently inefficient and places limits on the maximum speed of data transmission for a given modulation system. In the limit, if pulses were transmitted continuously, then the system would resemble a sinusoidal system. These issues are discussed in Chapter 5.

1.7   PENETRATION CHARACTERISTICS

One of the most important benefits of the UWB communication system that has
been raised is the ability of pulses to easily penetrate walls, doors, partitions, and
other objects in the home and office environment. In this section we will examine the
reported results for penetration of UWB pulses and comment on how these will affect
communication in the home and office.

Frequency f and wavelength λ are related by the speed of light c as is shown in
the following well-known equation:

$$\lambda \ [m] = \frac{c \ [\mathrm{m/s}]}{f \ [\mathrm{Hz}]} \qquad (1.4)$$

In other words, as the frequency increases the wavelength becomes shorter, and for
lower frequencies the wavelength is longer.

In conventional sinusoidal communication, lower frequency waves have the characteristic of being able to 'pass through' walls. doors, and windows because the length
of the wave is much longer than the material that it is passing through. On the other
hand, higher frequency waves will have more of their energy reflected from walls and
doors since their wavelength is much shorter.

UWB pulses are composed of a large range of frequencies, as is shown in Figure 1.2(b). One of the basic characteristics of early prototype UWB communication
systems was their ability to 'pass through walls', especially in comparison with IEEE
802.11 wireless local area network (WLAN) systems. The penetration capabilities of
UWB come only from the lower frequency components which were, for early systems,
mostly centered on 1 GHz. Since the 2002 ruling by the FCC (see Figure 1.7), the
center frequency for most UWB systems has substantially increased. This means that
the penetration characteristics of the signals have substantially decreased, especially
in comparison with the IEEE 802.11b systems which are centered at 2.4 GHz.

1.8   SPATIAL AND SPECTRAL CAPACITIES

Another basic property of UWB systems is their high *spatial capacity*, measured in
bits per second per square meter [bps/m^2] [11]. Spatial capacity is a relatively recent
term, and its use stems from the interest in even higher data rates, even over extremely
short distances.

Spatial capacity can be calculated as the maximum data rate of a system divided
by the area over which that system can transmit. The transmission area can be
calculated from the circular area, assuming a transmitter in the center; however, in
practice a rule of thumb is to use the square of the maximum transmission distance:

$$\text{Spatial capacity [bps/m}^2] = \frac{\text{Maximum data rate [bps]}}{\text{Transmission area [m}^2]} \qquad (1.5)$$

$$\text{Transmission area [m}^2] = \pi \times (\text{Transmission distance})^2 \qquad (1.6)$$

Table 1.2 Comparison of the spatial capacity of various indoor wireless systems.

System name	Maximum data rate [Mbps]	Transmission distance [m]	Spatial capacity [kbps/m^2]	Spectral capacity [bps/Hz]
UWB	100	10	318.3	0.013
IEEE 802.11a	54	50	6.9	2.7
Bluetooth	1	10	3.2	0.012
IEEE 802.11b	11	100	0.350	0.1317

For narrowband systems the most popular measure of capacity has been *spectral capacity*, measured in bits per second per hertz (bps/Hz), because the spectrum has been the most limited resource. Power has generally only been limited by safety and commercial reasons, such as the battery life of mobile devices:

$$\text{Spectral capacity [bps/Hz]} = \frac{\text{Maximum data rate [bps]}}{\text{Bandwidth [Hz]}} \qquad (1.7)$$

For UWB systems, which operate in other licensed spectra, the power has to be kept very low. This is compensated for by the use of extremely large bandwidths. Using the traditional measure of spectral capacity [bits/Hz], UWB has very low spectral capacity compared with existing systems. However, when comparing spatial capacity, UWB is extremely efficient. Table 1.2 shows a comparison of spatial and spectral capacity among various indoor wireless systems.

1.9 SPEED OF DATA TRANSMISSION

One of the advantages of UWB transmission for communications is its high data rate. While current chipsets are continually being improved, most UWB communication applications are targeting the range of 100–500 Mbps [12], which is roughly the equivalent of wired Ethernet to USB 2.0. It is significant that this data rate is 100 to 500 times the speed of Bluetooth, around 50 times the speed of the 802.11b, or 10 times the 802.11a WLAN standards.

As can be seen in Table 1.3 the current target data rate for indoor wireless UWB transmission is between 110 Mbps and 480 Mbps. This is fast compared with current wireless and wired standards. In fact, the speed of transmission is currently being standardized into three different speeds: 110 Mbps with a minimum transmission distance of 10 m; 200 Mbps with a minimum transmission distance of 4 m; and 480 Mbps with no fixed minimum distance.

The reasons for these particular distances lie mostly with different applications. For example, 10 m will cover an average room and may be suitable for wireless connectivity for home theater. A distance of less than 4 m will cover the distance

Table 1.3 Comparison of UWB bit rate with other wired and wireless standards.

Speed [Mbps]	Standard
480	UWB, USB 2.0
200	UWB (4 m minimum), 1394a (4.5 m)
110	UWB (10 m minimum)
90	Fast Ethernet
54	802.11a
20	802.11g
11	802.11b
10	Ethernet
1	Bluetooth

between appliances, such as a home server and a television. A distance of less than 1 m will cover the appliances around a personal computer.

1.10 COST

Among the most important advantages of UWB technology are those of low system complexity and low cost. UWB systems can be made nearly 'all-digital', with minimal radiofrequency (RF) or microwave electronics. The low component count leads to reduced cost, and smaller chip sizes invariably lead to low-cost systems. The simplest UWB transmitter could be assumed to be a pulse generator, a timing circuit, and an antenna.

However, as higher data rates are required, more complex timing circuitry is needed. To provide a multiple access system, additional complexity is required. Rake receivers add further circuitry, and the cost increases. Furthermore, chipset costs depend heavily on the number of units manufactured.

To reduce costs, during later product cycles more functionality is implemented on fewer chips, reducing die area and, thus, manufactured cost.

Therefore, at this early stage it is extremely difficult to quantify the cost of UWB communication systems. To take one early example, it has been reported [13] that the XtremeSpectrum chipset is priced at US$19.95 for 100000 units.

1.11 SIZE

The small size of UWB transmitters is a requirement for inclusion in today's consumer electronics. In the 802.15 working groups, consumer electronics companies have targeted the size of the wireless circuit to be small enough to fit into a Memory Stick or secure digital (SD) Card [12]. A chipset by XtremeSpectrum has a small size which enables compact flash implementation [13].

The main arguments for the small size of UWB transmitters and receivers are due to the reduction of passive components. However, antenna size and shape is another factor that needs to be considered. UWB antennas are considered in Chapter 7.

1.12 POWER CONSUMPTION

With proper engineering design the resultant power consumption of UWB can be quite low. As with any technology, power consumption is expected to decrease as more efficient circuits are designed and more signal processing is done on smaller chips at lower operating voltages.

The current target for power consumption of UWB chipsets is less than 100 mW. Table 1.4 shows some figures for power consumption of current chipsets [12].

Table 1.4 Power consumption of UWB and other mobile communication chipsets (LSI, large scale integration; RISC, reduced instruction set computer; MPU, microprocessor unit; TFT, thin film transistor).

Application chipset	Power consumption [mW]
802.11a	1500–2000
400 Mbps 1394 LSI	700
Mobile telephone RISC 32-bit MPU	200
Digital camera 12-bit A/D converter	150
UWB (target)	100
Mobile telephone TFT color display panel	75
MPEG-4 decoder LSI	50
Mobile telephone voice codec LSI	19

1.13 SUMMARY

In this chapter the basic properties of UWB signals were outlined, starting with the basic shape and spectrum of an UWB pulse. We saw that the power output and spectrum of UWB pulses are limited by regulation.

We showed that because UWB pulses are extremely short, with the consideration of fading, they can be filtered or ignored. They can readily be distinguished from unwanted multipath reflections because of the fine time resolution. This leads to the characteristic of multipath immunity.

The low-frequency components of the UWB pulses enable the signals to propagate effectively through materials, such as bricks and cement.

The large bandwidth of UWB systems means extremely high data rates can be achieved, and we showed that UWB systems have a potentially high spectral capacity.

Finally, we stated that UWB transmitters and receivers do not require expensive and large components, such as modulators, demodulators, and IF stages. This fact can reduce cost, size, weight, and power consumption of UWB systems compared with conventional narrowband communication systems.

Problems

Problem 1. Investigate the transmission power and bandwidth of three systems *not* shown in Table 1.1. (*Hint:* radio and television systems may differ significantly from those shown in the table due to differing local conditions.) Calculate their power spectral densities.

Problem 2. Implement the Matlab circuit of Figure 1.3, and confirm the output of the four scopes. Apply some modifications to the circuit if necessary.

Problem 3. Based on the Matlab circuit of Figure 1.3, implement a three-path multipath channel model. Add Gaussian noise. Show the output of the four scopes for two different values of delay of the second and third paths. Comment on the effect of multipath on the receiver and receiver design.

Problem 4. Write a Matlab program to output Figures 1.5(a) and 1.5(b).

Problem 5. Investigate the current FCC (or the regulator in your own country) regulations for UWB applications. What differences or similarities can you find with Figure 1.7?

Problem 6. Investigate and calculate the spatial and spectral capacities of three other wireless systems (outdoor wireless systems are acceptable) *not* shown in Table 1.2.

Problem 7. Investigate the reported speed of at least five current commercial UWB communication or laboratory prototypes. Plot a graph, with time as the horizontal axis and speed as the vertical axis.

Problem 8. Find the cost of five different wireless chipsets. Compare and contrast the complexity of the chipset, the maturity of the system, and the cost of the chipset. What, in your opinion, is the dominating factor for the cost of wireless chipsets?

Problem 9. For one standard or system (for example, UWB or IEEE standard 802.11a) find the cost of five commercial products. What fraction of the total product cost is contained in the wireless chipset?

2

Generation of UWB waveforms

One of the essential functions in communication systems is the representation of a message symbol by an analog waveform for transmission through a channel. As was shown in Chapter 1, in UWB systems the conventional analog waveform is a simple pulse that in general is directly radiated to the air. These short pulses have typical widths of less than 1 ns and thus a bandwidth of over 1 GHz.

In this chapter we will examine in detail how to generate pulse waveforms for UWB systems for simple cases of Gaussian wave shapes. We will discuss how to design pulses which meet requirements of spectral masks as mandated by government organizations. Finally, we will look at the practical constraints on pulse generation and the effects of imperfections on the pulse, and briefly discuss the effects of the channel on pulse shape.

2.1 INTRODUCTION

Sinusoidal electromagnetic waves have become so universal in radio communications that many people are not aware that the first communication systems were in fact pulse-based. Heinrich Hertz (1893) used a spark discharge to produce the electromagnetic waves for his experiment. These waves would be called colored noise today. Spark gaps and arc discharges between carbon electrodes were the dominant wave generators for about 20 years after Hertz's first experiments. Eventually, the development of rotating high-frequency generators and the electronic tube made the generation of sinusoidal currents and waves possible.

A strong incentive to use sinusoidal waves was provided by the need to operate several transmitters at the same time but to receive them selectively. This led to the

Ultra Wideband Signals and Systems in Communication Engineering Second Edition
M. Ghavami, L. B. Michael and R. Kohno © 2007 John Wiley & Sons, Ltd

development of transmitters and receivers on the basis of sinusoidal waves. Regulation followed common practice and led to the assignment of frequency bands for various radio services. However, consideration of textbooks published in the 1920s shows that nonsinusoidal waves were still used at that time.

Today's UWB systems employ nonsinusoidal wave shapes that should have certain properties when transmitted from the antenna. Emissions in UWB communication systems are constrained by the FCC regulation 47 CFR Section S15.5(d) [14], which states that

> Intentional radiators that produce class B emissions (damped wave) are prohibited.

Several nondamped waveforms have been proposed in the literature for UWB systems, such as Gaussian [15], Rayleigh, Laplacian, and cubic [16] waveforms, and modified Hermitian monocycles [17]. In all these waveforms the goal is to obtain a nearly flat frequency domain spectrum of the transmitted signal over the bandwidth of the pulse and to avoid a DC component. In order to understand the characteristics of different waveforms, we first discuss the theoretical definition of a damped wave.

2.1.1 Damped sine waves

What is a damped sine wave? According to [18] a damped sine wave is of the form

$$y_d(t) = Ae^{-\lambda t}\sin(2\pi f_0 t) \tag{2.1}$$

where A is an arbitrary amplitude, λ is the exponential decay coefficient, f_0 is the frequency of oscillation of the sine wave, and $t \geq 0$ is the time. Figure 2.1(a) demonstrates this waveform, and any wave of this general form can be considered as being a damped sine wave, or class B, emission.

Example 2.1

Calculate and sketch the time and frequency domain representations of Equation (2.1) for $A = 1$, $\lambda = 5 \times 10^9, 3 \times 10^9$, and $f = 5$ GHz.

Solution
Using the Fourier transform (Chapter 3) we can derive the frequency domain formula from the time domain representation as follows:

$$Y_d(f) = \mathcal{F}\{y_d(t)\}$$
$$= \int_0^\infty Ae^{-\lambda t}\sin(2\pi f_0 t)e^{-j2\pi ft}\,dt$$
$$= \frac{2\pi f_0 A}{\lambda^2 - 4\pi^2(f^2 - f_0^2) + j4\pi\lambda f} \tag{2.2}$$

where the final result of Equation (2.2) can be found by using a definite or indefinite table of integrals.

Substituting $A = 1$, $\lambda = 5 \times 10^9$, 3×10^9, and $f = 5$, the damped sine waves in the time domain are shown in Figure 2.1(a) and the frequency domain representation is given in Figure 2.1(b).

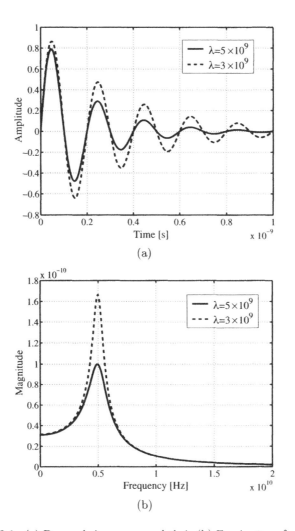

(a)

(b)

Fig. 2.1   (a) Damped sine waves and their (b) Fourier transforms.

The *effective bandwidth* W of a pulse is defined as

$$W = f_h - f_l \tag{2.3}$$

where f_l and f_h are the frequencies measured at nearest points with half of the maximum amplitude.

Example 2.2
For the damped sinusoids of Figure 2.1 calculate the effective bandwidth W in both cases. Comment on the relationship between W and the decay parameter λ.

Solution
The damped sinusoid of Figure 2.1 has a decay parameter $\lambda = 5 \times 10^9$. From Figure 2.1(b) we have $f_l = 3.25$ GHz and $f_h = 6.18$ GHz; hence, $W = 2.93$ GHz. When the decay parameter is $\lambda = 3 \times 10^9$, the effective bandwidth is $W = 1.69$ GHz.

   The value of W depends on the frequency of oscillation f_0 and the decay parameter λ. If λ decreases, W also decreases.

Figure 2.1(b) clearly indicates that the damping oscillations of the waveform have led to a small effective bandwidth and a sharp peak in the spectrum. This is in contradiction with the required flatness of the spectrum of permitted waveforms by the FCC [14]. Waveforms with damped oscillatory tails cannot be used in UWB communication systems because they can seriously interfere with existing communication systems that are available in the area. For this and several other reasons that will be discussed in this chapter, the topic of wave shaping is very important and should be considered in the design of practical UWB transmitters.

2.2 GAUSSIAN WAVEFORMS

A certain class of waveforms are called *Gaussian waveforms* because their mathematical definition is similar to the Gauss function. The zero-mean Gauss function is described by Equation (2.4), where σ is the standard deviation:

$$G(x) = \frac{1}{\sqrt{2\pi\sigma^2}} e^{-x^2/2\sigma^2} \tag{2.4}$$

The basis of these Gaussian waveforms is a *Gaussian pulse* represented by the following equation:

$$y_{g_1}(t) = K_1 e^{-(t/\tau)^2} \tag{2.5}$$

where $-\infty < t < \infty$, τ is the time-scaling factor, and K_1 is a constant. More waveforms can be created by a sort of high-pass filtering of this Gaussian pulse. Filtering acts in a manner similar to taking the derivative of Equation (2.5). For example, a *Gaussian monocycle*, the first derivative of a Gaussian pulse, has the form

$$y_{g_2}(t) = K_2 \frac{-2t}{\tau^2} e^{-(t/\tau)^2} \tag{2.6}$$

where $-\infty < t < \infty$ and K_2 is a constant. A Gaussian monocycle has a single zero crossing. Further derivatives yield additional zero crossings, one additional zero crossing for each additional derivative. If the value of τ is fixed, by taking an additional derivative, the fractional bandwidth decreases, while the centre frequency increases.

A *Gaussian doublet* is the second derivative of Equation (2.5) and is defined by

$$y_{g_3}(t) = K_3 \frac{-2}{\tau^2} \left(1 - \frac{2t^2}{\tau^2} \right) e^{-(t/\tau)^2} \tag{2.7}$$

where $-\infty < t < \infty$ and K_3 is a constant. Figures 2.2(a) and 2.2(b) show a Gaussian pulse, a Gaussian monocycle, and a Gaussian doublet in the time and frequency domains, respectively. In all three cases, $\tau = 50$ ps is assumed. The constants K_1, K_2, and K_3 are added to Equations (2.5)–(2.7) in order to determine the energy of the UWB Gaussian pulses. If E_1, E_2, and E_3 are the required energies of y_{g_1}, y_{g_2}, and y_{g_3}, respectively, it is possible to calculate the relation between them in a straightforward manner. The energy of $y_{g_1}(t)$ is defined and computed as

$$E_1 = \int_{-\infty}^{\infty} y_{g_1}(t)^2 \, dt$$

$$= \int_{-\infty}^{\infty} K_1^2 e^{-2(t/\tau)^2} \, dt$$

$$= K_1^2 \tau \sqrt{\pi/2} \tag{2.8}$$

Hence, the resulting formula for K_1 is as follows:

$$K_1 = \sqrt{\frac{E_1}{\tau \sqrt{\pi/2}}} \tag{2.9}$$

Employing an analogous procedure for K_2 and K_3 yields

$$K_2 = \sqrt{\frac{\tau E_2}{\sqrt{\pi/2}}} \tag{2.10}$$

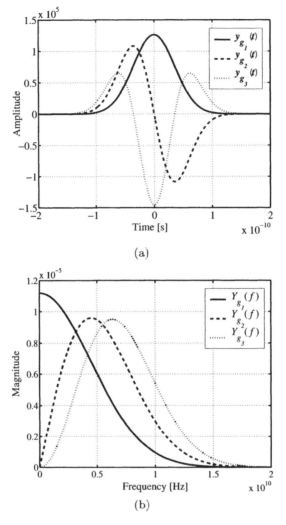

Fig. 2.2 A Gaussian pulse, monocycle, and doublet in the (a) time and (b) frequency domains. The Gaussian pulse has a large DC component.

and

$$K_3 = \tau \sqrt{\frac{\tau E_3}{3\sqrt{\pi/2}}} \tag{2.11}$$

Pulses of Figure 2.2(a) are all plotted for unit energy. The curves illustrated in Figure 2.2(b) are obtained by taking the Fourier transforms of Equations (2.5)–(2.7).

For example, the frequency domain representation of $y_{g_1}(t)$ can be derived as follows:

$$Y_{g_1}(f) = \mathcal{F}\{y_{g_1}(t)\}$$

$$= \int_{-\infty}^{\infty} K_1 e^{-(t/\tau)^2} e^{-j2\pi ft}\, dt$$

$$= K_1 \tau \sqrt{\pi}\, e^{-(\pi\tau f)^2} \tag{2.12}$$

Since $y_{g_2}(t)$ and $y_{g_3}(t)$ are proportional to the first and second derivatives of $y_{g_1}(t)$, respectively, the Fourier transforms of $y_{g_2}(t)$ and $y_{g_3}(t)$ are the Fourier transforms of $y_{g_1}(t)$ multiplied by $(K_2/K_1)(j2\pi f)$ and $(K_3/K_1)(j2\pi f)^2$, respectively:

$$Y_{g_2}(f) = K_2 \tau \sqrt{\pi}(j2\pi f)e^{-(\pi\tau f)^2} \tag{2.13}$$

$$Y_{g_3}(f) = K_3 \tau \sqrt{\pi}(j2\pi f)^2 e^{-(\pi\tau f)^2} \tag{2.14}$$

The effective bandwidths of the Gaussian wave shapes of Figure 2.2 are now calculated. For the Gaussian pulse $y_{g_1}(t)$, which is actually a low-pass waveform, $f_l = 0$ and f_h is calculated by inserting $Y_{g_1}(f) = K_1 \tau \sqrt{\pi}/2$ into Equation (2.12). The result is $0.265/\tau$; hence,

$$W_{g_1} = 5.3 \text{ GHz} \tag{2.15}$$

For the Gaussian monocycle defined in Equation (2.6) and shown in Figure 2.2 we can calculate $f_l = 1.44$ GHz and $f_h = 8.65$ GHz, which yields

$$W_{g_2} = 7.21 \text{ GHz} \tag{2.16}$$

For the Gaussian doublet defined in Equation (2.7) and shown in Figure 2.2 we can find $f_l = 3.07$ GHz and $f_h = 10.42$ GHz, which gives

$$W_{g_3} = 7.35 \text{ GHz} \tag{2.17}$$

The choice of which Gaussian waveform to use is usually driven by system design and application requirements. In any case the waveform should have a center frequency of about 3 to 10 GHz. All these three Gaussian waveforms have a dramatically different spectrum from a damped sine wave. One important characteristic of these waveforms is that they are almost uniformly distributed over their frequency spectrum and, therefore, are noise-like.

2.3 DESIGNING WAVEFORMS FOR SPECIFIC SPECTRAL MASKS

It has already been explained in Chapter 1 that UWB signals are like noise. Transmitted energy is spread over a very wide bandwidth and the energy or power density of signals is very low. In this section we investigate another interesting tool for designing different pulse shapes. The pulses introduced in this section have deep spectral nulls which are at the same frequency as conventional narrowband communication systems. The result of these nulls is the extensive reduction of interference caused by our UWB system on nearby receivers.

2.3.1   Introduction

Single short pulse (or impulse) generation is the traditional and fundamental approach for generating UWB waveforms. By varying the pulse characteristics, the characteristics of the energy in the frequency spectrum may be defined based on the desired design criteria. Generally, there are three parameters of interest when defining the properties of energy which is filling a specified frequency spectrum:

- the intended bandwidth of the transmitted energy should be carefully defined and considered;

- we have to limit the available energy to within the specified UWB frequency spectrum;

- the generated energy should be defined so as to center within the spectrum of interest (that is, the *center frequency*).

Pulse duration in the time domain determines the bandwidth in the frequency domain. As a rule of thumb we may write

$$\frac{1}{\text{Duration}} \approx \text{Bandwidth} \tag{2.18}$$

Pulse repetition is a characteristic that may determine the center frequency of a band of transmitted energy if the repetition is regular.

Pulse shape determines the characteristics of how the energy occupies the frequency domain.

Since the highest spectral efficiency is one of the most important objectives for UWB communication, as much bandwidth as is practical should be used to take advantage of the capacity made available when bandwidth is very large.

As with traditional radio architectures, data can be modulated on the pulses in a number of ways, including amplitude, phase, time position, or any combination of these. Single-pulse architectures offer relatively simple radio designs, but provide little flexibility where spectrum management is an objective. Examples of scenarios where managing the spectrum might be desirable are:

- matching different regulatory requirements in different international regions;

- dynamically sensing interfering technologies and suspending usage of contending frequencies;

- choosing to use narrower bands of spectrum to either share spectrum in a local area or to enable lower cost devices that do not require large bandwidth for a specific application.

Another area where managing the spectrum might be desirable is in performance management. Increasing the performance of existing designs may require the complete redesign of a radio for higher performance implementations, forgoing backward compatibility with earlier implementations [19].

2.3.2   Multiband modulation

Multiband modulation [20] is another approach to modulating information with wide-band technology. Multiband UWB (MW-UWB) modulation provides a method where, for example, the 7.5 GHz of spectrum made available to UWB technology by the FCC (see Chapter 1) may be split into multiple smaller frequency bands. This could be accomplished by choosing to implement a few large bands or many small bands, all of which would be stacked across the legally available frequency spectrum.

As stated previously, UWB technology uses pulses to modulate information over a wide band of frequencies. Pulse shape is the primary characteristic that determines the distribution of energy within the frequency domain, and properly shaping the pulse will concentrate more of the energy in the center lobe of the energy band, reducing side lobe energy and reducing chances for adjacent band interference. To effectively fill the specified spectrum, multiple frequency bands of energy must be generated with different center frequencies and must be spaced across the spectrum. Center frequency selection is accomplished when using a pseudo-carrier oscillation to generate and shape the required UWB pulse. The frequency of the pseudo-carrier oscillation determines the center frequency of the band, while the outline of the oscillation defines the pulse shape and, thus, defines the bandwidth.

The idea of multiband modulation is to use multiple frequency bands to efficiently utilize the UWB spectrum by transmitting multiple UWB signals at the same time. The signals do not interfere with each other because they operate at different frequencies. Assume that the Gaussian pulse of Figure 2.2(a) is modulated by a sinusoidal signal to form a transmitted pulse as follows:

$$
\begin{aligned}
s_j(t) &= \sqrt{2}\, y_{g_1}(t) \cos(2\pi f_j t) \\
&= \sqrt{2}\, K_1 e^{-(t/\tau)^2} \cos(2\pi f_j t)
\end{aligned}
\tag{2.19}
$$

where f_j is the modulating frequency, which is not much higher than the bandwidth of the Gaussian pulse. The waveform defined by Equation (2.19) is shown in Figure 2.3(a) for $f_j = 4$ GHz, $\tau = 0.5$ ns, and unit energy. The Fourier transform of Equation (2.19) can be calculated as follows:

$$
\begin{aligned}
S_j(f) &= \frac{\sqrt{2}}{2} [Y_{g_1}(f - f_j) + Y_{g_1}(f + f_j)] \\
&= \frac{\sqrt{2\pi}}{2} K_1 \tau \left[e^{-(\pi\tau(f-f_j))^2} + e^{-(\pi\tau(f+f_j))^2} \right]
\end{aligned}
\tag{2.20}
$$

The spectrum of the modulated Gaussian pulse for positive frequencies is illustrated in Figure 2.3(b) and extends around 4 GHz. Based on the discussion in Section 2.1, this kind of wave shape is of damped sinusoidal type and normally is not to be used for UWB communications. However, if a combination of several pulses like $s_j(t)$ contribute to the construction of the transmitted signal the result can be quite satisfactory.

As an example of multiband modulation consider the case where five different frequencies, $f_1 = 4$, $f_2 = 5$, $f_3 = 6$, $f_4 = 7$, and $f_5 = 8$ GHz, are utilized for

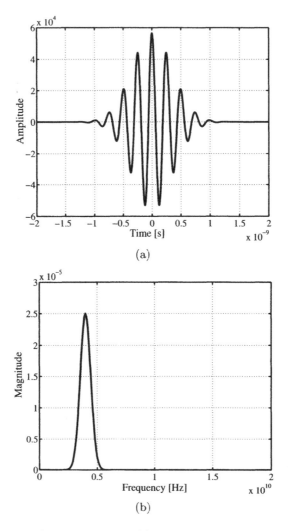

Fig. 2.3 (a) A modulated Gaussian pulse and (b) its frequency domain presentation. The centre frequency is 4 GHz.

modulating Gaussian pulses, and each of these signals is transmitted simultaneously to achieve a very high bit rate or used as a means of multiple access to allow multiple users to communicate at the same time. In addition, several standard digital modulation techniques can also be used on each individual UWB signal. The output of the modulated UWB signals can be added together before transmission. At the receiver the modulated UWB signals must be separated before demodulation. A combination of five modulated Gaussian pulses can be simply done as follows:

$$s_{mb}(t) = \sum_{j=1}^{5} s_j(t)$$

$$= \sqrt{2}\,K_1 e^{-(t/\tau)^2} \sum_{j=1}^{5} \cos\left(2\pi f_j t\right) \qquad (2.21)$$

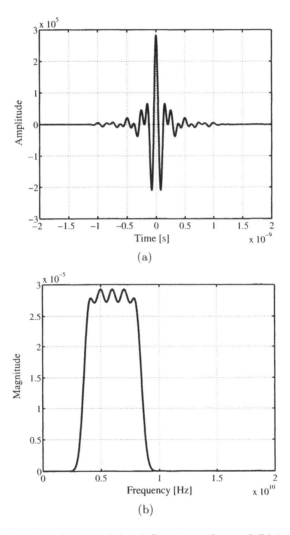

Fig. 2.4 (a) A combination of five modulated Gaussian pulses and (b) its frequency domain presentation.

A plot of Equation (2.21) is shown in Figure 2.4(a) for $\tau = 0.5$ ns, $E_1 = 1$, and $K_1 = \sqrt{E_1/(\tau\sqrt{\pi/2})} = 39\,947$. By using Equations (2.20) and (2.21), the spectrum of $s_{mb}(t)$ is easily calculated as follows:

$$S_{mb}(f) = \sum_{j=1}^{5} S_j(f)$$

$$= \frac{\sqrt{2\pi}}{2} K_1 \tau \sum_{j=1}^{5} \left[e^{-(\pi\tau(f-f_j))^2} + e^{-(\pi\tau(f+f_j))^2} \right] \qquad (2.22)$$

The result is shown in Figure 2.4(b). It is obvious from this figure that a wide frequency range from about 4 to about 8 GHz is covered by the spectrum, and therefore the damping sinusoidal properties of Figure 2.4(b) are removed.

There are several advantages resulting from using MB-UWB modulation. One advantage is the ability to efficiently use the entire 7.5 GHz of spectrum made available by the FCC. Choosing appropriate multiband widths can ensure full usage of the entire spectrum, as compared with single-pulse systems. Figure 2.4(b) shows how multiple bands of wideband energy might efficiently fill the specified spectrum. Notice the rapid attenuation at the band edges.

Another compelling advantage is that separate bands stacked across the available spectrum may be treated independently, creating a new level of flexibility for UWB. Coexistence and interference are other interesting areas that could benefit from a multiband approach. Any UWB system will be expected to share the approved spectrum with existing technologies. It is important first to protect these existing technologies (UWB interfering with narrowband) using mitigation techniques. Overlay concepts must then be developed to ensure UWB is robust (narrowband interfering with UWB). It will be the responsibility of UWB developers to avoid interference with existing spectrum users if UWB is to be successful.

Some form of interference mitigation will also be possible by multiband implementation. The most obvious technology that will require coexistence next to UWB is WLAN in the 5 GHz spectrum. A multiband implementation would be able to identify potential interference, and either reduce power in the contending band or turn it off completely.

Let us consider again the wave shape of Figure 2.4(a) that was given by the time domain equation (Equation (2.21)) and remove the term corresponding to 5 GHz. This term belongs to f_2. The resulting pulse shape is illustrated in Figure 2.5(a). Although the removal of a frequency component might be hard to recognize when comparing Figures 2.4(a) and 2.5(a), the frequency response of Figure 2.5(b) is quite self-explanatory. As is clear from this figure, the wave shape designed, based on the multiband modulation technique, has no 5 GHz component to interfere with other communication systems operating in this frequency.

The level of the null produced in Figure 2.5(b) can be adjusted adaptively by changing the number of bands and the parameter of the Gaussian pulse used for wave shaping. The following example will consider a modification of parameters in order to increase the depth of the desired null in the frequency spectrum.

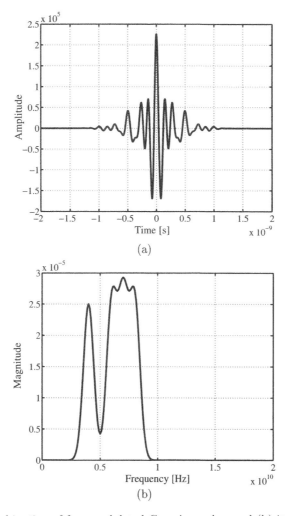

Fig. 2.5 (a) A combination of four modulated Gaussian pulses and (b) its frequency domain presentation after removing the 5 GHz band for interference mitigation.

Example 2.3
Modify the parameters of Figure 2.5 in such a way that the result presents a deeper null at the 5 GHz frequency band.

Solution
Let us modify the numerical values of the parameters of the example of Figure 2.5 as follows:

$$f_1, f_2, \ldots, f_{13} = 3.6, 3.9, 4.2, 4.5, 5.5, 5.8, 6.1,$$
$$6.4, 6.7, 7, 7.3, 7.6, 7.9 \text{ GHz} \qquad (2.23)$$

and $\tau = 1.5$ ns. The individual pulses that form the combined UWB wave are
narrower this time and they are arranged with a smaller frequency distance.
Moreover, there is no frequency component at 5 GHz. In fact, the two nearest
frequency components to 5 GHz are $f_4 = 4.5$ GHz and $f_5 = 5.5$ GHz. The
resulting pulse shape and its spectrum are illustrated in Figures 2.6(a) and 2.6(b),
respectively. We note that, compared with Figure 2.5(b), the null produced
in the frequency characteristic is much deeper, and consequently the potential
interfering behavior of this UWB signal on 5 GHz communication systems will
be considerably reduced.

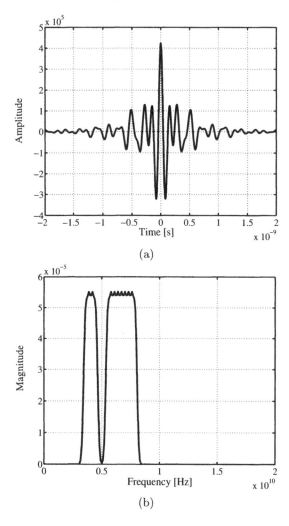

Fig. 2.6 The deeper null produced by changing the number of bands and the parameter of the
Gaussian pulse used for wave shaping. (a) The pulse shape and (b) its spectrum.

Another way of interference mitigation with the assistance of wave shaping can be done by using a delay and sum combination of several multiband modulated pulses. As an example, consider the following signal generated at the transmitter side:

$$s'_{mb}(t) = \sum_{j=1,3,4,5} s_j\,(t - t_j)$$
$$= \sqrt{2}\,K_1 \sum_{j=1,3,4,5} e^{-(t-t_j/\tau)^2} \cos\left(2\pi f_j(t - t_j)\right) \qquad (2.24)$$

where $f_1 = 4$, $f_3 = 6$, $f_4 = 7$, and $f_5 = 8$ GHz are utilized for modulating the Gaussian pulses, $\tau = 0.5$ ns, and $K_1 = \sqrt{1/(\tau\sqrt{\pi/2})} = 39\,947$. The delay times are chosen as $t_1 = 0$, $t_3 = 4$, $t_4 = 6$, and $t_5 = 8$ ns. Figure 2.7(a) shows the time domain representation of the pulses of Equation (2.24), and Figure 2.7(b) demonstrates the frequency spectrum of these pulses. Again, it is clear that the wave shape designed this way has no 5 GHz component capable of interfering with other communication systems operating in this frequency.

2.4   PRACTICAL CONSTRAINTS AND EFFECTS OF IMPERFECTIONS

The utilization of the RF spectrum based on band sharing has long been discussed in different publications [21]. Instead of allocating the spectrum by frequency division we can construct radio waveforms from a set of orthogonal functions of inherent broad bandwidth and, in effect, allocate the spectrum by a scheme that is equivalent to code division. In this section we attempt to identify the key technical issues that govern the acceptability of the pulse-shaping scheme in RF transmission. Some practical difficulties of implementing such a scheme for most applications are presented.

The main argument regarding the practical constraint of UWB pulse shapes is that the generation, transmission, and reception of UWB radio waves are less natural than sinusoids in most applications. Three reasons can be mentioned to support this idea [22]:

1. Radiation efficiency varies rapidly with frequency. Some UWB waveforms have considerable low-frequency content. It is a difficult practical problem to transfer useful low-frequency energy to a radiator and to accomplish its emission as radio waves, particularly if directivity is desired.

2. Relatively high-power switching circuits that might be essential in transmitters and receivers of UWB systems are usually not attractive in practice. In addition, the complexity and time-varying behavior of pulse-shaping circuits could possibly be a drawback regarding cost and reliability compared with conventional transmission systems.

3. The physical medium through which radio waves travel is time-variant on the scale of typical radio-wave travel times and linear for normal field strength, but it is dispersive in general. Consequently, whereas sinusoids after having travelled

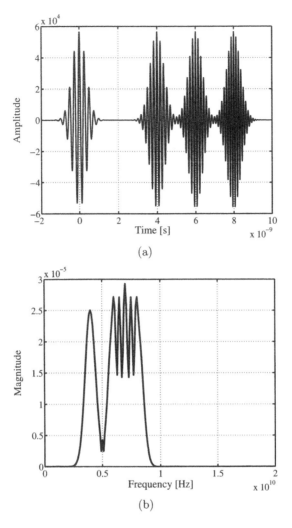

Fig. 2.7 (a) A combination of four delayed modulated Gaussian pulses and (b) its frequency domain presentation after removing the 5 GHz band for interference mitigation.

through the channel are still almost sinusoids, complex functions of time no longer keep their exact shapes and, particularly, their orthogonality.

2.5 SUMMARY

In this chapter we discussed the various aspects and features of pulse generation in UWB communication systems. We started with the definition of damped sinusoids that are not appropriate for UWB systems due to spectral energy peaks and the possibility of interfering with other communication systems. A class of waveforms called Gaussian pulses, monocycles, and doublets was defined based on the Gauss function and differentiation. Several properties of these functions, such as energy, frequency spectrum, and bandwidth, were explained. The concept of a Gaussian pulse train for time modulation of UWB signals was also discussed. The effect of changing the pulse duration and repetition rate was examined.

Designing waveforms for specific spectral masks using multiband modulation was discussed. We noticed that, by changing various parameters of the modulation (for example, the number of frequency components and the scaling parameter), we can create controlled deep nulls in the spectrum of the pulse shape and, hence, reduce the potential interfering behavior of the UWB system.

Finally, practical constraints and the effects of imperfections were briefly mentioned. We observed that radiation efficiency is a very rapidly varying characteristic with frequency. Moreover, the problem with the necessity of complicated electronic circuits was explained. We also noted that the time variations and dispersive behavior of the medium through which the radio wave travels might change the shape of waveforms and, consequently, assumptions such as the orthogonality of the pulses might no longer be maintained.

Problems

Problem 1. In Section 2.2 we defined and investigated the properties of three basic Gaussian waveforms $y_{g_1}(t)$, $y_{g_2}(t)$, and $y_{g_3}(t)$ in Equations (2.5), (2.6), and (2.7), respectively. Proceed in the same way to derive $y_{g_4}(t)$ and:

(i) find its time domain equation;

(ii) calculate its frequency domain equation;

(iii) compare these characteristics with other Gaussian waveforms;

(iv) find the constant K_4; and

(v) calculate the bandwidth W_{g_4}.

Problem 2. Consider multiband modulation using modulated Gaussian pulses, such as indicated in Equation (2.21). By using 10 different frequencies and a proper value for the parameter τ, design a pulse shape that produces a deep null at the frequency of 7 GHz.

3

Signal-processing techniques for UWB systems

The basic tools used in the systematic way of solving problems of science and engineering usually include an appropriate mathematical formulation, a suitable mathematical solution, and, finally, a physical interpretation. In conventional narrowband signals and systems only the steady-state solutions to a given physical problem are considered. When the system is excited by a wideband signal the problem becomes very difficult.

It should be clearly stated that any analysis involving UWB signals and systems does not require new mathematical techniques. The treatment of systems excited by wideband waveforms can be adequately done by using, for example, two-sided Laplace transforms, which have been available in the mathematical literature for over 200 years. Moreover, time domain analysis results contain equivalent information to frequency domain results, if an adequate interpretation of these results is performed. Relative or fractional bandwidth is totally irrelevant to the analysis, as Fourier or Laplace transforms are completely general in nature and require no such bandwidth restrictions.

3.1 THE EFFECTS OF A LOSSY MEDIUM ON A UWB TRANSMITTED SIGNAL

Analytical solution of the transient behavior of a class of microwave networks whose impulse response could be modeled with a train of impulses began in 1962, and was called time domain electromagnetics [7]. Early works demonstrated that time domain waveforms could be used to obtain wideband frequency domain information of two-port networks at 1 GHz. Later investigations presented a new method utilizing reflected and transmitted waves to obtain the complex permittivity and permeability of linear materials over a broad range of microwave frequencies.

In this section we do not intend to consider a detailed and sophisticated examination of the issues related to UWB electromagnetics. In fact, Chapter 4 is devoted to UWB propagation and channel modeling. The example mentioned here explains the problem of UWB pulse propagation in a lossy medium [2].

Maxwell's equations predict the propagation of electromagnetic energy away from time-varying sources (current and charge) in the form of waves. As shown in Figure 3.1, consider a linear, homogeneous, isotropic medium characterized by (μ, ϵ, σ) in a source-free region (sources in region 1, source-free environment is region 2). Assume that a UWB signal propagates in this medium that has an effective lossy permittivity of ϵ and a conductivity of σ. The magnetic permeability of the medium is μ. The target of the problem is to calculate the transient electric and magnetic fields if the excitation at the source is an impulsive signal, such as a UWB pulse.

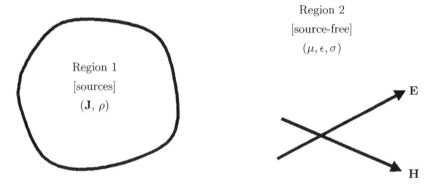

Fig. 3.1 Regions including the source and lossy medium for calculations of the electric and magnetic fields of a UWB signal.

If the electric and magnetic fields in region 2 are a function of time t and length z, then it can be shown that the electric and magnetic fields $E(z,t)$ and $H(z,t)$ satisfy the following equations:

$$\mu\epsilon\frac{\partial^2 E(z,t)}{\partial t^2} + \mu\sigma\frac{\partial E(z,t)}{\partial t} - \frac{\partial^2 E(z,t)}{\partial z^2} = 0 \tag{3.1}$$

$$\mu\epsilon\frac{\partial^2 H(z,t)}{\partial t^2} + \mu\sigma\frac{\partial H(z,t)}{\partial t} - \frac{\partial^2 H(z,t)}{\partial z^2} = 0 \tag{3.2}$$

where z represents the direction of propagation and t is time. These equations are called instantaneous wave equations. The properties of an electromagnetic wave (direction of propagation, velocity of propagation, wavelength, frequency, attenuation, etc.) can be determined by examining the solutions to the wave equations that define the electric and magnetic fields of the wave.

The solution of this problem is similar to the solution of pulse propagation in a lossy transmission line. For example, if R, L, G, and C are the resistance, inductance, conductance, and capacitance per unit of a lossy transmission line, respectively, then the voltage $v(z,t)$ on the transmission line is given by

$$LC\frac{\partial^2 v(z,t)}{\partial t^2} + (RC + LG)\frac{\partial v(z,t)}{\partial t} + RGv(z,t) - \frac{\partial^2 v(z,t)}{\partial z^2} = 0 \qquad (3.3)$$

where the transmission line is assumed to be oriented in the z-direction. A solution to this problem has been derived using the Laplace transform. The solution for the current on the transmission line satisfies a similar differential equation to the one that was satisfied by the electric field. The original solution, first developed more than 35 years ago, is given as

$$v(z,t) = \left(e^{-\rho z/\nu}\delta\left[t - \frac{z}{\nu}\right] + \frac{\eta z}{\nu\chi}e^{-\rho t}I_1(\eta\chi)\right) \cdot u\left[t - \frac{z}{\nu}\right] \qquad (3.4)$$

$$i(z,t) = \left(e^{-\rho z/\nu}\delta\left[t - \frac{z}{\nu}\right] + e^{-\rho t}\left(\frac{\eta t}{\chi}I_1(\eta\chi) - \eta I_0(\eta\chi)\right)\right) \cdot u\left[t - \frac{z}{\nu}\right] \qquad (3.5)$$

where $u[\cdot]$ is the unit step function and $\delta[\cdot]$ is the delta function. The functions defined by I_0 and I_1 are the modified Bessel functions of the first kind, with order zero and one, respectively. Also

$$\rho = \frac{1}{2}\left(\frac{R}{L} + \frac{G}{C}\right) \qquad (3.6)$$

$$\nu = \frac{1}{\sqrt{LC}} \qquad (3.7)$$

$$\eta = \frac{1}{2}\left(\frac{R}{L} - \frac{G}{C}\right) \qquad (3.8)$$

$$\chi = \sqrt{t^2 - \frac{z^2}{\nu^2}} \qquad (3.9)$$

Because of the existence of the step function in the above equations, the solution is guaranteed to be causal.

Now, propagation of a pulse in a lossy medium is exactly similar to pulse propagation in a lossy transmission line. Therefore, setting $R = 0$ in the above expression leads to the following equivalent:

$$v(z,t) \Rightarrow E(z,t)$$
$$i(z,t) \Rightarrow H(z,t)$$
$$L \Rightarrow \mu$$
$$C \Rightarrow \epsilon$$
$$G \Rightarrow \sigma$$

The electric and magnetic fields propagating in a lossy medium due to an arbitrarily shaped waveform excitation can easily be obtained by convolving the above expressions with the excitation waveform. It can be proven that electric and magnetic

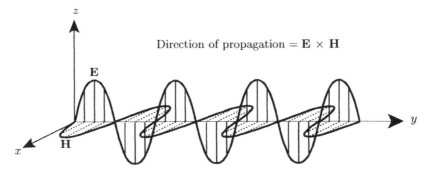

Fig. 3.2   Propagation of electric and magnetic fields.

fields lie in a plane perpendicular to the direction of propagation and to each other. Figure 3.2 shows the relationship between **E** and **H** for the previously assumed uniform plane wave propagating in a slightly lossy medium.

Although this section may appear to bear little relevance to UWB signal processing it is important to consider the effects of transmission medium, such as antennas and physical propagation channel. The above analysis presents the fundamentals required for such effects to be analyzed and appreciated.

3.2   TIME DOMAIN ANALYSIS

The broadband characteristics of UWB communication systems can be obtained using either the time domain or the frequency domain. Each of these methods has its own advantages and disadvantages. In time domain analysis an important objective is to find the impulse response of a system. For example, the system can be a propagation channel, a transmitter antenna, or a receiver antenna. In this section various features and issues related to time domain processing of signals (in particular, UWB signals) are introduced.

3.2.1   Classification of signals

Electric signals can be classified in different ways [23]. Some examples of these classifications are now explained.

3.2.1.1   Continuous time and discrete time signals   By the term *continuous time* signal we mean a real or complex function of time $s(t)$, where the independent variable t is continuous. If t is a discrete variable (i.e. $s(t)$ is defined at discrete times), then the signal $s(t)$ is a *discrete time* signal. A discrete time signal is often identified as a sequence of numbers, denoted by $s(n)$, where n is an integer. For example, the functions

$$s(t) = (t^2 - 1)e^{-t^2/4} \tag{3.10}$$

and

$$s(n) = (n^2 - 1)e^{-n^2/4} \tag{3.11}$$

can represent two UWB pulses in the continuous and discrete time domains, respectively. A plot of these functions is illustrated in Figure 3.3.

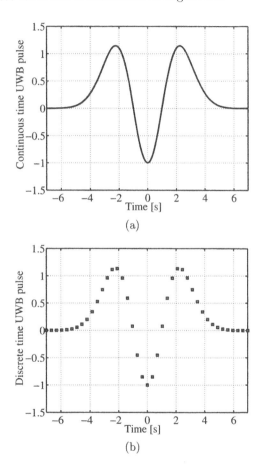

Fig. 3.3 Examples of (a) continuous time and (b) discrete time UWB signals.

3.2.1.2 Analog and digital signals If a continuous time signal $s(t)$ can take on any values in a continuous time interval, then $s(t)$ is called an *analog* signal. If a discrete time signal can take on only a finite number of distinct values $s(n)$, then the signal is called a *digital* signal.

Since UWB pulses are extremely short, it is difficult to use conventional analog-to-digital (A/D) conversion with a single converter. Further investigation into different techniques that could be used to acquire the analog sub-nanosecond-pulsed signal that comes out from a UWB front end is required.

3.2.1.3 Deterministic and random signals
Deterministic signals are those signals whose values are completely specified for any given time. For example, e^{-t^2} is well defined for any value of time.

Random signals are those signals that take random values at any given time. For instance, the noise at the receiver is a random signal and must be modeled probabilistically.

3.2.1.4 Periodic and nonperiodic signals
A signal $s(t)$ is called a *periodic signal* if

$$s(t) = s(t + nT_0) \tag{3.12}$$

where T_0 is the *period* and the integer $n > 0$. If $s(t) \neq s(t + T_0)$ for all t and any T_0, then $s(t)$ is a *nonperiodic* or *aperiodic* signal.

3.2.1.5 Power and energy signals
A *complex* signal $s(t)$ is a *power* signal if the average normalized power P is finite, where

$$0 < P = \lim_{T \to \infty} \frac{1}{T} \int_{-T/2}^{T/2} s(t)s^*(t) \, dt < \infty \tag{3.13}$$

and $s^*(t)$ is the complex conjugate of $s(t)$.

A *complex* signal $s(t)$ is an *energy* signal if the normalized energy E is finite, where

$$0 < E = \int_{-\infty}^{\infty} s(t)s^*(t) \, dt$$

$$= \int_{-\infty}^{\infty} |s(t)|^2 \, dt < \infty \tag{3.14}$$

Energy signals have finite energy. Power signals have finite and nonzero power. In fact, any signal with finite energy will have zero power and any signal with nonzero power will have infinite energy.

3.2.2 Some useful functions

For the time domain analysis of a system, special functions are defined. These functions, although not necessarily practical, can be approximated and employed for investigating the time domain characteristics of an unknown system.

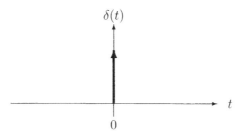

Fig. 3.4   The unit impulse function $\delta(t)$.

3.2.2.1 *Unit impulse function*   The *unit impulse function*, also known as the *Dirac delta function* $\delta(t)$, is shown in Figure 3.4 and is defined by

$$\int_{-\infty}^{\infty} s(\lambda)\delta(\lambda - t)\, d\lambda = s(t) \tag{3.15}$$

An alternative definition is

$$\int_{-\infty}^{\infty} \delta(t)\, dt = 1 \tag{3.16}$$

and

$$\delta(t) = \begin{cases} \infty, & t = 0 \\ 0, & t \neq 0 \end{cases} \tag{3.17}$$

The response of a system to a unit impulse function is called the *impulse response* of the system and is an important measure of its time domain characteristics.

3.2.2.2 *Unit step function*   The *unit step function* $u(t)$ is shown in Figure 3.5 and is defined according to

$$u(t) = \begin{cases} 1, & t > 0 \\ 0, & t < 0 \end{cases} \tag{3.18}$$

This is related to the unit impulse function $\delta(t)$ by

$$u(t) = \int_{-\infty}^{t} \delta(\lambda)\, d\lambda \tag{3.19}$$

and

$$\frac{du(t)}{dt} = \delta(t) \tag{3.20}$$

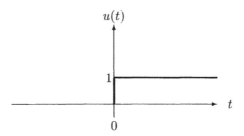

Fig. 3.5   The unit step function $u(t)$.

3.2.2.3 Sinc function   A *sinc function* is denoted by

$$\text{sinc}(t) = \frac{\sin \pi t}{\pi t} \tag{3.21}$$

and is illustrated in Figure 3.6.

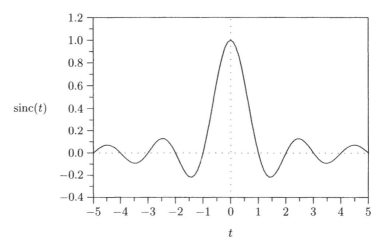

Fig. 3.6   The sinc function.

3.2.2.4 Rectangular function   A single *rectangular pulse* is denoted by

$$\Pi\left(\frac{t}{T}\right) = \begin{cases} 1, & |t| < \dfrac{T}{2} \\[2mm] 0, & |t| > \dfrac{T}{2} \end{cases} \tag{3.22}$$

3.2.2.5 Triangular function   A *triangular function* is denoted by

$$\Lambda\left(\frac{t}{T}\right) = \begin{cases} 1 - \dfrac{|t|}{T}, & |t| < T \\ 0, & |t| > T \end{cases} \tag{3.23}$$

3.2.3 Some useful operations

3.2.3.1 Time average   The *time average operator* is given by

$$\langle[\cdot]\rangle = \lim_{T\to\infty} \frac{1}{T} \int_{-T/2}^{T/2} [\cdot]\, dt \tag{3.24}$$

If a waveform is periodic, the time average operator can be reduced to the following:

$$\langle[\cdot]\rangle = \frac{1}{T_0} \int_{a-T_0/2}^{a+T_0/2} [\cdot]\, dt \tag{3.25}$$

where T_0 is the period of the waveform and a is an arbitrary real constant, which may be taken to be zero. Equation (3.25) readily follows from Equation (3.24) because integrals over successive time intervals that are T_0 seconds wide have identical area if the waveform is periodic. As these integrals are summed, the total area and T are proportionally larger, resulting in a value for the time average that is the same as just integrating over one period and dividing by T_0. In summary, Equation (3.24) may be used to evaluate the time average of any type of waveform. Equation (3.25) is valid only for periodic waveforms.

3.2.3.2 Direct current value   The *direct current* (DC) value of a waveform is given by

$$\langle s(t)\rangle = \lim_{T\to\infty} \frac{1}{T} \int_{-T/2}^{T/2} s(t)\, dt \tag{3.26}$$

We can see that this is the time average of $s(t)$. Over a finite interval of interest, the DC value is

$$\langle s(t)\rangle = \frac{1}{t_2 - t_1} \int_{t_1}^{t_2} s(t)\, dt \tag{3.27}$$

3.2.3.3 Power and energy   The *instantaneous power* (incremental work divided by incremental time) is given by

$$p(t) = v(t)i(t) \tag{3.28}$$

where $v(t)$ denotes voltage and $i(t)$ denotes current.

The *average power* is given by

$$\begin{aligned} P &= \langle p(t)\rangle \\ &= \langle v(t)i(t)\rangle \end{aligned} \tag{3.29}$$

The *root mean square* (rms) value of $s(t)$ is given by

$$S_{\text{rms}} = \sqrt{\langle s^2(t) \rangle} \qquad (3.30)$$

If a load is resistive, the average power is given by

$$P = \frac{\langle v^2(t) \rangle}{R} = \langle i^2(t) \rangle R$$

$$= \frac{V_{\text{rms}}^2}{R} = I_{\text{rms}}^2 R = V_{\text{rms}} I_{\text{rms}} \qquad (3.31)$$

where R is the value of the resistive load. When $R = 1\,\Omega$, P becomes the *normalized power*.

The *average normalized power* of a *real-valued* signal $s(t)$ is given by

$$P = \langle s^2(t) \rangle$$

$$= \lim_{T \to \infty} \frac{1}{T} \int_{-T/2}^{T/2} s^2(t)\,dt \qquad (3.32)$$

The *total normalized energy* of a *real-valued* signal $s(t)$ is given by

$$E = \lim_{T \to \infty} \int_{-T/2}^{T/2} s^2(t)\,dt \qquad (3.33)$$

3.2.3.4 Decibel   The decibel gain of a circuit is given by

$$\text{dB} = 10 \log_{10}\left(\frac{P_{\text{out}}}{P_{\text{in}}}\right) \qquad (3.34)$$

If resistive loads are involved, it can be reduced to

$$\text{dB} = 20 \log_{10}\left(\frac{V_{\text{rms out}}}{V_{\text{rms in}}}\right) + 10 \log_{10}\left(\frac{R_{\text{in}}}{R_{\text{load}}}\right) \qquad (3.35)$$

or

$$\text{dB} = 20 \log_{10}\left(\frac{I_{\text{rms out}}}{I_{\text{rms in}}}\right) + 10 \log_{10}\left(\frac{R_{\text{load}}}{R_{\text{in}}}\right) \qquad (3.36)$$

If normalized powers are used,

$$\text{dB} = 20 \log_{10}\left(\frac{V_{\text{rms out}}}{V_{\text{rms in}}}\right) = 20 \log_{10}\left(\frac{I_{\text{rms out}}}{I_{\text{rms in}}}\right) \qquad (3.37)$$

The *decibel power level* with respect to 1 mW is given by

$$\text{dBm} = 10 \log_{10}\left(\frac{\text{Actual power level in watts}}{10^{-3}}\right) \qquad (3.38)$$

or, alternatively, the *decibel power level* with respect to 1 W is given by

$$dBW = 10 \log_{10}(\text{actual power level in watts})$$

The *decibel voltage level* with respect to a 1 mV rms level is given by

$$dBmV = 10 \log_{10} \left(\frac{V_{\text{rms}}}{10^{-3}} \right) \tag{3.39}$$

3.2.3.5 Cross-correlation The *cross-correlation* of two *real-valued power* waveforms $s_1(t)$ and $s_2(t)$ is defined by

$$R_{12}(\tau) = \langle s_1(t)s_2(t+\tau) \rangle$$

$$= \lim_{T \to \infty} \frac{1}{T} \int_{-T/2}^{T/2} s_1(t)s_2(t+\tau)\, dt \tag{3.40}$$

If $s_1(t)$ and $s_2(t)$ are periodic with the same period T_0, then

$$R_{12}(\tau) = \frac{1}{T_0} \int_{-T_0/2}^{T_0/2} s_1(t)s_2(t+\tau)\, dt \tag{3.41}$$

The cross-correlation of two *real-valued energy* waveforms $s_1(t)$ and $s_2(t)$ is defined by

$$R_{12}(\tau) = \int_{-\infty}^{\infty} s_1(t)s_2(t+\tau)\, dt \tag{3.42}$$

Correlation is a useful operation to measure the similarity between two waveforms. To compute the correlation between waveforms it is necessary to specify which waveform is being shifted. In general, $R_{12}(t)$ is not equal to $R_{21}(t)$, where $R_{21}(t) = \langle s_2(t)s_1(t+\tau) \rangle$.

The cross-correlation of two complex waveforms is $R_{12}(\tau) = \langle s_1^*(t)s_2(t+\tau) \rangle$.

3.2.3.6 Autocorrelation The *autocorrelation* of a *real-valued power* waveform $s_1(t)$ is defined by

$$R_{11}(\tau) = \langle s_1(t)s_1(t+\tau) \rangle$$

$$= \lim_{T \to \infty} \frac{1}{T} \int_{-T/2}^{T/2} s_1(t)s_1(t+\tau)\, dt \tag{3.43}$$

If $s_1(t)$ is *periodic* with fundamental period T_0, then

$$R_{11}(\tau) = \frac{1}{T_0} \int_{-T_0/2}^{T_0/2} s_1(t)s_1(t+\tau)\, dt \tag{3.44}$$

The autocorrelation of a *real-valued energy* waveform $s_1(t)$ is defined by

$$R_{11}(\tau) = \int_{-\infty}^{\infty} s_1(t)s_1(t+\tau)\, dt \tag{3.45}$$

The autocorrelation of a *complex power* waveform is $R_{11}(\tau) = \langle s_1^*(t)s_1(t+\tau) \rangle$.

3.2.3.7 Convolution   The *convolution* of a waveform $s_1(t)$ with a waveform $s_2(t)$ is given by

$$s_3(t) = s_1(t) * s_2(t)$$

$$= \int_{-\infty}^{\infty} s_1(\lambda) s_2(t - \lambda) \, d\lambda$$

$$= \int_{-\infty}^{\infty} s_1(\lambda) s_2(-(\lambda - t)) \, d\lambda \tag{3.46}$$

where $*$ denotes the convolution operation. The third line of this equation is obtained by:

1. time reversal of $s_2(t)$ to obtain $s_2(-\lambda)$;

2. time shifting of $s_2(-\lambda)$ to obtain $s_2(-(\lambda - t))$;

3. multiplying $s_1(\lambda)$ and $s_2(-(\lambda - t))$ to form the integrand $s_1(\lambda) s_2(-(\lambda - t))$.

As an example, let us convolve a rectangular waveform $s_1(t)$ with an exponential waveform $s_2(t)$, which are defined by the following equations:

$$s_1(t) = \begin{cases} 1, & 0 < t < T \\ 0, & \text{elsewhere} \end{cases} \tag{3.47}$$

$$s_2(t) = e^{-t/T} u(t) \tag{3.48}$$

The steps involved in the convolution are illustrated in Figure 3.7.

3.2.4   Classification of systems

A system is a mathematical model that relates the output signal to the input signal of a physical process. Representation of a system is shown in Figure 3.8. Classification of signals and systems will help us in finding a suitable mathematical model for a given physical process that is to be analyzed.

3.2.4.1 Linear and nonlinear systems   Let $x_i(t)$ and $y_i(t)$, $i \geq 1$, be input and output signals of a system, respectively. As shown in Figure 3.9, a system is called a *linear* system if the input $x_1(t) + x_2(t) + \cdots + x_i(t) + \cdots$ produces a response $y_1(t) + y_2(t) + \cdots + y_i(t) + \cdots$ and $ax_i(t)$ produces $ay_i(t)$ for all input signals $x_i(t)$ and scalar a. This is known as the *superposition theorem*, and a linear system obeys this principle.

In practice, it may be found that a system is only linear over a limited range of input signals.

A *nonlinear* system does not obey the superposition theorem.

3.2.4.2 Causal and noncausal systems   Let $x(t)$ and $y(t)$ be the input and output signals of a system (Figure 3.10). A *causal* (physically realizable) system produces an

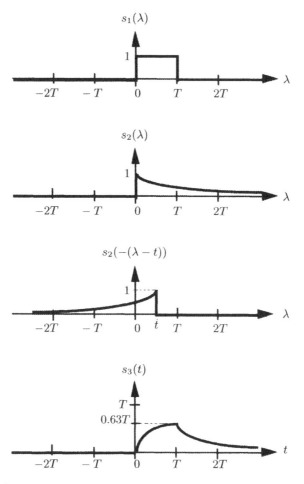

Fig. 3.7 Convolution of a rectangular waveform with an exponential waveform.

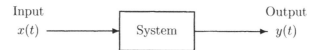

Fig. 3.8 Representation of a system with input $x(t)$ and output $y(t)$.

output response at time t_1 for an input at time t_0, where $t_0 \leq t_1$. In other words, a causal system is one whose response does not begin before the input signal is applied.

A *noncausal* system response will begin before the input signal is applied. It can be made realizable by introducing a positive time delay into the system.

3.2.4.3 Time-invariant and time-varying systems If the input $x(t - t_0)$ produces a response $y(t - t_0)$, where t_0 is any real constant, the system is called a *time-invariant*

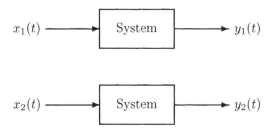

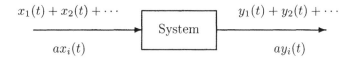

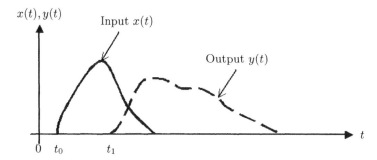

Fig. 3.9   Conditions for a system to be linear.

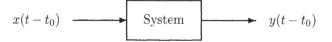

Fig. 3.10   Signals associated with a causal system.

system (Figure 3.11). If the above condition is not satisfied, the system is called a *time-varying* system.

A system is called a *linear time-invariant* (LTI) system if the system is linear and time-invariant.

Fig. 3.11   Time-invariant system.

3.2.5 Impulse response

The *impulse response* $h(t)$ of an LTI system is defined as the response of the system when the input signal $x(t)$ is a delta function $\delta(t)$. The output $y(t)$ of an LTI system can be expressed as the convolution of the input signal $x(t)$ and the impulse response $h(t)$ of the system, i.e.

$$y(t) = x(t) * h(t) = \int_{-\infty}^{\infty} x(\lambda)h(t - \lambda)\,d\lambda \qquad (3.49)$$

$$y(t) = h(t) * x(t) = \int_{-\infty}^{\infty} h(\lambda)x(t - \lambda)\,d\lambda \qquad (3.50)$$

For a causal LTI system

$$y(t) = \int_{0}^{\infty} x(\lambda)h(t - \lambda)\,d\lambda = \int_{0}^{\infty} h(\lambda)x(t - \lambda)\,d\lambda \qquad (3.51)$$

3.2.6 Distortionless transmission

In communication systems, distortionless transmission is often desired. This implies that the output signal $y(t)$ is given by

$$y(t) = Kx(t - t_d) \qquad (3.52)$$

where K is a constant and t_d is a time delay.

3.3 FREQUENCY DOMAIN TECHNIQUES

Frequency response is the gain and phase response of a circuit or other unit under test at all frequencies of interest. Although the formal definition of frequency response includes both the gain and phase, in common usage the frequency response often only implies the magnitude or gain.

The frequency response is directly related to the Fourier transform of a continuous time or discrete time signal.

3.3.1 Fourier transforms

The *Fourier transform* of a continuous time function $h(t)$ is denoted by $H(f)$ and is defined as follows:

$$H(f) = \mathcal{F}\{h(t)\} = \int_{-\infty}^{\infty} h(t)e^{-j2\pi ft}\,dt \qquad (3.53)$$

The *discrete time Fourier transform* (DTFT) of a sequence $h(n)$ is given by

$$H(e^{j\Omega}) = \sum_{n=-\infty}^{\infty} h(n)e^{-j\Omega n} \qquad (3.54)$$

if the sum converges. Uniform convergence is assured for all sequences that are absolutely summable,

$$\sum_{n=-\infty}^{\infty} |h(n)| < \infty \tag{3.55}$$

Note that Equation (3.54) corresponds to the definition of the two-sided z-transform, which will be defined later. Moreover, from its definition it follows directly that the DTFT is periodic in Ω with period 2π,

$$H(e^{j(\Omega+k2\pi)}) = \sum_{n=-\infty}^{\infty} h(n)e^{-j(\Omega+k2\pi)n}$$

$$= \sum_{n=-\infty}^{\infty} h(n)e^{-j\Omega n} = H(e^{j\Omega}) \tag{3.56}$$

This implies that one period defines the DTFT completely.

Frequency response measurements require the excitation of the system under test at all relevant frequencies. The fastest way to perform the measurement is to use a broadband excitation signal that excites all frequencies simultaneously and use fast Fourier transform (FFT) techniques to measure at all of these frequencies at the same time. Noise and nonlinearity are best minimized by using random noise excitation, but short impulses or rapid sweeps (chirps) may also be used. When the desired resolution bandwidth of interest is low the fastest way to measure the frequency response functions is to use FFT-based techniques.

3.3.2   Frequency response approaches

There are various approaches for practical measurement of the frequency response of a system. Some of them are explained now.

3.3.2.1 Sine generator/voltmeter We apply a sine wave to the input of the system under test and measure the output voltage. Then, we repeat this process for each frequency. The gain of the system is the ratio of the output voltage to the input voltage.

Another variation of this method is that we apply a sine wave to the input and measure the phase of the output relative to the input at each frequency of interest. This method has the advantage of being low cost and simple. It is also quite slow, and the following assumptions must be fulfilled in order for the measurement to be accurate:

1. The output voltage of the signal generator is stable during the measurement and also at all frequencies. If there is doubt about this the voltage must be measured at each frequency.

2. The system does not create significant distortion.

3. There is no significant noise on the output of the system. Otherwise, the measured output voltage will be too high. As a rule of thumb, if there is 1% distortion or noise in the system the error will be of the same order of magnitude.

4. The output must be statistically correlated to the input. This assumption is normally true in high-fidelity analog systems. However, in systems with complex transmission mechanisms and/or with digital encoding, echo cancelling, and other adaptive techniques, this assumption may not be fulfilled.

To account for all of the above, you can use digital signal-processing techniques including FFT and cross spectral methods.

Another variation that we might use is a swept sine wave generator and an associated voltmeter. The requirements of this method are as follows:

1. The sweep time for a given bandwidth must be greater than the reciprocal of the desired bandwidth. For example, if a resolution of 1 MHz is desired, the sweep time for the 1 MHz must be at least 1 µs.

2. The integration time of the voltmeter must be short enough, otherwise it cannot respond fully.

This is a variant of the first method in that it uses continuous swept sine waves, instead of discretely stepped sine waves. This variation can be faster than the first technique, but it must fulfill the same assumptions.

Certain instruments may have 'adaptive sweep', where the sweep rate adapts to the rate of change of the output signal. For example, when sweeping through a very sharp resonance, the sweep rate is reduced to fully resolve the resonance peak.

In the third variation a swept sine with tracking filtering similar to that of the second variation is used, but this method has the advantage of being able to reject noise and distortion from the system by using a filter on the output that follows the frequency of the input. This method must meet all the requirements regarding averaging times.

A sophisticated variant of this method offsets (delays) the receiving filter with a fixed frequency offset (corresponding to a fixed time delay) and makes it possible to measure the frequency response of delayed signal paths.

When making this measurement we make sure that the output impedance of the sine generator is low compared with the input impedance of the unit under test. Otherwise, the actual input voltage applied may drop or be changed as a function of frequency.

3.3.2.2 *Transient or noise excitation with cross spectral techniques* It might be possible to use any signal that contains frequency components in the range of interest. The signals are not required to have the same amplitude. However, all measurements using cross spectral techniques require simultaneous measurement of both input and output signals, using simultaneously sampling A/D converters.

The frequency response can be computed as

$$H_{xy}(f) = \frac{G_{xy}}{G_{xx}} \tag{3.57}$$

where G_{xy} is the cross spectrum and G_{xx} is the autospectrum of the input.

This technique computes the correlation between the input and output signals (as a function of frequency) and, hence, rejects noise and distortion. The greater the number of statistical samples that are included in the averaging, the greater the noise and distortion rejection and, hence, the greater the accuracy of the measurement. The resulting statistical function, called the *cross spectrum*, is then normalized for the actual amplitude of the signal at each frequency on the input (called the *autospectrum*, or, more commonly, the averaged spectrum). This gives the frequency response function, which contains both magnitude and phase information. The magnitude is typically shown on a logarithmic Y axis (in dB), and the phase is often shown on a $0°$ to $360°$ scale.

This approach has the advantage of overcoming noise, distortion, and noncorrelated effects. It also corrects for any loading effects on the input to the system. In addition, the technique can be extremely rapid, because it measures all frequencies of interest simultaneously. Its only weakness is that its SNR can be lower than the swept sine with the tracking filter technique.

3.3.2.3 Naturally occurring excitation Sometimes, it is not possible to insert an excitation signal into the system to be tested. However, if you want to measure the frequency response function you can use the naturally occurring 'input signals' coming from the surrounding environment as excitation signals. Using cross spectral techniques you can measure the input signal and cross-correlate it with the output signal.

When making this measurement you should take extreme care to account for triggering and windowing conditions, and also consider the potential time delays between the input and output. Thus, this technique is only recommended for experienced professionals with a thorough understanding of digital signal-processing techniques.

3.3.3 Transfer function

In the frequency domain the Fourier transform of

$$y(t) = h(t) * x(t) \tag{3.58}$$

is

$$Y(f) = H(f)X(f) \tag{3.59}$$

where $H(f)$ is the Fourier transform of $h(t)$. The function $H(f)$ is called the *transfer function* or *frequency response* of the LTI system.

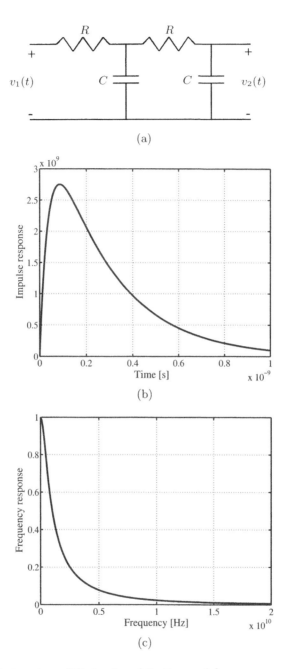

Fig. 3.12   A simple two-stage RC circuit and its time and frequency response for $R = 10\ \Omega$ and $C = 10$ pF.

Example 3.1
Figure 3.12(a) shows a two-stage RC circuit with two resistors R and two capacitors C. Find and sketch the impulse response and frequency response of the circuit.

Solution
Using well-known circuit theory techniques we can derive the differential equation relating $v_1(t)$ and $v_2(t)$ as follows:

$$R^2C^2\ddot{v}_2(t) + 3RC\dot{v}_2(t) + v_2(t) = v_1(t) \tag{3.60}$$

The impulse response of the circuit is defined as $h(t) = v_2(t)$ if $v_1(t) = \delta(t)$. It can be shown that a solution in the form of

$$v_2(t) = \left(k_0 + k_1 e^{-(3+\sqrt{5})t/2RC} + k_2 e^{-(3-\sqrt{5})t/2RC} \right) u(t) \tag{3.61}$$

can satisfy the differential equation. We also note that, because the right-hand side of Equation (3.60) has an infinite discontinuity of type $\delta(t)$ at $t=0$, the left-hand side should have the same level of discontinuity at the term involving the second derivative. Consequently, the first derivative of $v_2(t)$ will have a finite discontinuity at $t=0$ and, hence, $v_2(t)$ should be continuous. Assuming zero initial charge at $t=0$ on both capacitors C_1 and C_2, we should have $v_2(0)=0$, and hence

$$v_2(t) = \left(-(k_1+k_2) + k_1 e^{-(3+\sqrt{5})t/2RC} + k_2 e^{-(3-\sqrt{5})t/2RC} \right) u(t) \tag{3.62}$$

Substituting Equation (3.62) into Equation (3.60) and equating both sides of the resulting expression yields

$$k_2 = -k_1 = \frac{1}{RC\sqrt{5}} \tag{3.63}$$

Therefore, the impulse response of the circuit becomes

$$h(t) = v_2(t) = \frac{1}{RC\sqrt{5}} \left(-e^{-(3+\sqrt{5})t/2RC} + e^{-(3-\sqrt{5})t/2RC} \right) u(t) \tag{3.64}$$

as is shown in Figure 3.12(b) for $R = 10\ \Omega$ and $C = 10$ pF.

The frequency response or the transfer function of the circuit can be calculated by finding the Fourier transform of the impulse response. First, let us consider the Fourier transform of $h_1(t) = e^{\alpha t}u(t)$. Using Equation (3.53) we can write

$$H_1(f) = \int_{-\infty}^{\infty} e^{\alpha t} u(t) e^{-j2\pi ft} \, dt$$

$$= \int_0^{\infty} e^{(\alpha - j2\pi f)t} \, dt$$

$$= \frac{1}{j2\pi f - \alpha} \tag{3.65}$$

Now the Fourier transform of Equation (3.64) can be derived as follows:

$$H(f) = \frac{1}{RC\sqrt{5}} \left[-\frac{1}{j2\pi f + (3 + \sqrt{5})/2RC} + \frac{1}{j2\pi f + (3 - \sqrt{5})/2RC} \right]$$

$$= \frac{1}{1 + j6\pi RCf - 4\pi^2 R^2 C^2 f^2} \tag{3.66}$$

An illustration of this equation is shown in Figure 3.12(c) and indicates the low-pass characteristic of the circuit.

3.3.4 Laplace transform

The one-sided Laplace transform of the function $h(t)$ is defined by

$$H(s) = \int_0^{\infty} h(t) e^{-st} \, dt \tag{3.67}$$

where t is real and $s = \sigma + j\omega$ is complex. There is also a two-sided Laplace transform obtained by setting the lower integration limit from 0 to $-\infty$. Since we will be analyzing only causal linear systems using the Laplace transform we can use either. However, it is customary in engineering treatments to use the one-sided definition. The one- and two-sided Laplace transforms are also called the unilateral and bilateral Laplace transforms, respectively.

When evaluated along the $s = j\omega$ axis (i.e. $\sigma = 0$), the Laplace transform reduces to the Fourier transform. The Laplace transform can therefore be viewed as a generalization of the Fourier transform from the real line (a simple frequency axis) to the complex plane. One benefit of the more general Laplace transform is the ability to transform signals which have no Fourier transform. To see this we can write the Laplace transform as

$$H(s) = \int_0^{\infty} h(t) e^{-(\sigma + j\omega)t} \, dt$$

$$= \int_0^{\infty} [h(t) e^{-\sigma t}] e^{-j\omega t} \, dt \tag{3.68}$$

We can thus interpret the Laplace transform as the Fourier transform of an exponentially enveloped input signal. For $\sigma > 0$ (the so-called 'right-half-plane'), this exponential weighting forces the Fourier-transformed signal toward zero as $t \to \infty$. As long as the signal $h(t)$ does not increase faster than e^{Bt} for some B, its Laplace transform will exist for all $\sigma > B$.

Example 3.2
Find the Laplace transform of the output signal $v_2(t)$ in Example 3.1.

Solution
Similar to the Fourier transform, we might write

$$H_1(s) = \int_0^\infty e^{\alpha t} e^{-st}\, dt$$

$$= \frac{1}{s - \alpha} \tag{3.69}$$

Now $V_2(s)$ will be easily calculated as follows:

$$V_2(s) = \frac{1}{RC\sqrt{5}} \left[-\frac{1}{s + (3 + \sqrt{5})/2RC} + \frac{1}{s + (3 - \sqrt{5})/2RC} \right]$$

$$= \frac{1}{1 + 3RCs + R^2 C^2 s^2} \tag{3.70}$$

3.3.5 z-transform

The z-transform, like the Laplace transform, is an indispensable mathematical tool for the design, analysis, and monitoring of systems. The z-transform is the discrete time counterpart of the Laplace transform and a generalization of the Fourier transform of a sampled signal. Like the Laplace transform, the z-transform allows insight into the transient behavior, the steady-state behavior, and the stability of discrete time systems. A working knowledge of the z-transform is essential to the study of UWB signals and systems.

The *bilateral z-transform* of the discrete time signal $h(n)$ is defined to be

$$H(z) = \sum_{n=-\infty}^{\infty} h(n) z^{-n} \tag{3.71}$$

where z is a complex variable. Since signals are typically defined to begin (become nonzero) at time $n = 0$ and since all filters are assumed to be causal, the lower summation limit given above may be written as 0 rather than $-\infty$ to yield the *unilateral z-transform*.

As mentioned earlier, the DTFT corresponds to the definition of the two-sided z-transform for the special case of $z = e^{j\Omega}$. The region of convergence for the z-transform is determined by the range of z, where the sequence $h(n)z^{-n}$ is absolutely summable.

The z-transform of a signal $h(n)$ can be regarded as a polynomial in z^{-1}, with coefficients given by the signal samples. As an example, the finite duration signal

$$h = [\ldots, 0, 0, 1, 2, 3, 0, 0, \ldots] \tag{3.72}$$

has the z-transform

$$H(z) = 1 + 2z^{-1} + 3z^{-2} \tag{3.73}$$

Mathematically, the entire signal is converted to a complex scalar indexed by z.

The z-transform of a signal h will always exist provided:

1. the signal starts at a finite time; and

2. it is *asymptotically exponentially bounded,* i.e. there exists a finite integer n_f and finite real numbers $A \geq 0$ and σ, such that $|h(n)| < Ae^{\sigma n}$ for all $n \geq n_f$. The bounding exponential may be growing with $n(\sigma > 0)$.

These are not the most general conditions for existence of the z-transform, but they suffice for our purposes here.

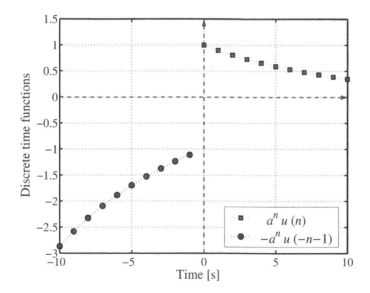

Fig. 3.13 Two discrete time exponential functions $h_1(n)$ and $h_2(n)$ for $a = 0.9$.

Example 3.3

Find the z-transform of the following functions:

$$h_1(n) = a^n u(n) \tag{3.74}$$

$$h_2(n) = -a^n u(-n-1) \tag{3.75}$$

These discrete time functions are plotted in Figure 3.13 for $a = 0.9$. Discuss the convergence and stability of both exponential functions.

Solution

Using the definition of z-transform we can write

$$
\begin{aligned}
H_1(z) &= \sum_{n=-\infty}^{\infty} a^n u(n) z^{-n} \\
&= \sum_{n=0}^{\infty} a^n z^{-n} \\
&= \sum_{n=0}^{\infty} (az^{-1})^n
\end{aligned}
\tag{3.76}
$$

This series converges only if

$$\sum_{n=0}^{\infty} |az^{-1}|^n < \infty$$

or, equivalently, if $|az^{-1}| < 1$ or $|z| > a$. In this case we have

$$H_1(z) = \frac{1}{1 - az^{-1}} = \frac{z}{z - a} \tag{3.77}$$

For $h_2(n)$ the z-transform can be written as

$$
\begin{aligned}
H_2(z) &= \sum_{n=-\infty}^{\infty} [-a^n u(-n-1)] z^{-n} \\
&= - \sum_{n=-\infty}^{-1} a^n z^{-n} \\
&= - \sum_{n=1}^{\infty} a^{-n} z^{n} \\
&= 1 - \sum_{n=0}^{\infty} (a^{-1}z)^n
\end{aligned}
\tag{3.78}
$$

This series converges only if $|z| < a$. In this case we have

$$H_2(z) = 1 - \frac{1}{1 - a^{-1}z} = \frac{1}{1 - az^{-1}} = \frac{z}{z - a} \tag{3.79}$$

It is interesting to note that both $h_1(n)$ and $h_2(n)$ have identical z-transforms although they are right-sided and left-sided in time, respectively. The convergence region of a z-transform is very important for its computation.

References [24] and [25] provide a detailed discussion of analog and digital signal processing, respectively.

3.3.6   The relationship between the Laplace transform, the Fourier transform, and the z-transform

The Laplace transform, the Fourier transform, and the z-transform are closely related in that they all employ complex exponentials as their basis function. For right-sided signals (zero-valued for negative time index) the Laplace transform is a generalization of the Fourier transform of a continuous time signal and the z-transform is a generalization of the Fourier transform of a discrete time signal. In the previous section we have shown that the z-transform can be derived as the Laplace transform of a discrete time signal. In the following we explore the relation between the z-transform and the Fourier transform. Using the relationship

$$z = e^s = e^\sigma e^{j\omega} = re^{j2\pi f} \tag{3.80}$$

where $s = \sigma + j\omega$ and $\omega = 2\pi f$, we can rewrite the z-transform in the following form:

$$H(z) = \sum_{n=-\infty}^{+\infty} h(n)r^{-n}e^{-j2\pi n f} \tag{3.81}$$

Note that when $r = e^\sigma = 1$ the z-transform becomes the Fourier transform of a sampled signal given by

$$H(z = e^{j2\pi f}) = \sum_{n=-\infty}^{+\infty} h(n)e^{-j2\pi n f} \tag{3.82}$$

Therefore, the z-transform is a simple generalization of the Fourier transform of a sampled signal. Like the Laplace transform, the basis functions for the z-transform are damped or growing sinusoids of the form $z^{-n} = r^{-n}e^{-j2\pi n f}$. These signals are particularly suitable for transient signal analysis, such as UWB signals. Fourier basis functions are steady complex exponentials, $e^{-j2\pi n f}$, of time-invariant amplitudes and phases, and are suitable for steady-state or time-invariant signal analysis.

A similar relationship exists between the Laplace transform and the Fourier transform of a continuous time signal. The Laplace transform is a one-sided transform with the lower limit of integration at $t = 0$, whereas the Fourier transform, Equation (3.53), is a two-sided transform with the lower limit of integration at $t = -\infty$. However, for a one-sided signal that is zero-valued for $t < 0$ the limits of integration for the Laplace and the Fourier transforms are identical. In that case, if the variable s in the Laplace transform is replaced with the frequency variable $j2\pi$, then the Laplace integral becomes the Fourier integral. Hence, for a one-sided signal the Fourier transform is a special case of the Laplace transform corresponding to $s = j2\pi f$ and $\sigma = 0$.

3.4   UWB SIGNAL-PROCESSING ISSUES AND ALGORITHMS

The distinctive properties of UWB signals require modification of analytical methods, signal-processing methods, and new constructional solutions compared with conventional narrowband signals which merely require a description of envelope and phase for a full characterization. For UWB systems, representation of the signal with a single envelope and phase is neither convenient nor possible.

A UWB signal is typically composed of a train of sub-nanosecond pulses, resulting in a bandwidth over 1 GHz. Since the total power is spread over such a wide range of frequencies, its PSD is extremely low. This minimizes the interference caused to existing services that already use the same spectrum. On account of the large bandwidth used, UWB links are capable of transmitting data over tens of megabits per second. Other benefits include a low probability of interference, precise location capability, and the possibility of transceiver implementation using simple architectures. It is mostly desired that UWB systems can be implemented using digital architectures because they bring low cost, ease of design, and flexibility. A single receiver would be able to support different bit rates, quality of service, and operating ranges. The vision of fully configurable software radios is an exciting one, and it appears UWB may be more amenable to such a realization than conventional, narrowband systems.

Figure 3.14 shows a generic block diagram for a digital UWB receiver [26]. A key component of such a system is the analog-to-digital converter (ADC). The ADC sampling rate for digitizing a UWB signal must be on the order of a few gigasamples per second (GSPS). Even with the most modern process technologies this constitutes a serious challenge. Most reported data converters operating at this speed employ interleaving [27], with each channel typically based on a FLASH converter. The latter is the architecture of choice for high-speed designs, but is not suitable for high-resolution applications. An N-bit FLASH converter uses $2N$ comparators, so its power and area scale exponentially with resolution. Among recently reported high-speed ADCs (>1 GSPS) representing the state of the art, none has a resolution exceeding 8 bits. The minimum number of bits needed for reliable detection of a UWB signal is, therefore, a critical parameter. If excessively large it can render an all-digital receiver infeasible.

Information in such a system is typically transmitted using a collection of narrow pulses with a very low duty cycle of about 1%. The duty cycle is the ratio of pulse duration to pulse period. Each user is assigned a different PN sequence that is used to encode the pulses in either position (PPM) or polarity (binary phase shift keying

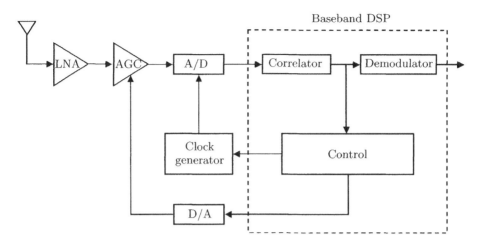

Fig. 3.14 Block diagram of a simple digital UWB receiver (LNA, low noise amplifier; AGC, automatic gain control; DSP, digital signal processing).

(BPSK), also known as bi-phase modulation (BPM)). Channelization (the use of a single wideband and high-capacity facility to create many relatively narrowband and lower capacity channels by subdividing the wideband facility) is thus based on the assigned code, as in the case of code division multiple access (CDMA) systems.

Suppose the bitstream is denoted by a sequence of binary symbols b_j (with values $+1$ or -1) for $j = -\infty, \ldots, \infty$. A single bit is represented using N_c pulses, where N_c refers to the length of the PN code c_i. For BPSK the code modulates the polarity of a pulse within each frame. For PPM it modulates the pulse positions (incrementing or decrementing them by multiples of T_c). Data modulation is achieved by setting the sign of the block of N_c pulses for BPSK. For PPM we append an additional time shift τ_{b_j} whose value depends on whether b_j is $+1$ or -1. Each frame has duration T_f; the duration of each bit is thus given by $N_c T_f$. Letting A denote the amplitude of each pulse $p(t)$, the transmitted signal $s(t)$ can be written as follows for the two different modulation schemes:

$$s_{\mathrm{BPSK}}(t) = A \sum_{j=-\infty}^{\infty} \sum_{i=0}^{N_c-1} b_j c_i p(t - jN_c T_f - iT_f) \qquad (3.83)$$

and

$$s_{\mathrm{PPM}}(t) = A \sum_{j=-\infty}^{\infty} \sum_{i=0}^{N_c-1} p(t - jN_c T_f - iT_f - c_i T_c - \tau b_j) \qquad (3.84)$$

As mentioned above, each bit is represented by N_c pulses. This redundancy is one component of the signal's processing gain (PG). The other component is the duty cycle, a ratio of the short duration of each pulse to the large interval between successive ones. Processing gain refers to the boost in effective SNR as a UWB signal is processed by a correlating receiver:

$$\text{PG [dB]} = 10\log(N_c) + 10\log\left(\frac{1}{\text{duty cycle}}\right) \tag{3.85}$$

Optimal detection of a noisy signal is based on matched filtering. This would entail correlating the received signal $r(t)$ against a template $s(t)$ that is an exact replica of the original transmitted signal and then feeding the correlator output to a slicer. However, generating exact replicas of sub-nanosecond pulses is a difficult problem and is highly susceptible to timing jitter.

A more tractable approach is to use a template signal comprising a train of rectangular pulses that are, in general, wider than the actual received pulses, but are coded with the same PN sequence.

Figure 3.15 shows the structure of the received and template signals, whereas Figure 3.16 depicts the operations necessary for demodulation. The signals $s_0(t)$ and $s_1(t)$ represent the template signals corresponding to a transmitted 0 and 1, respectively. They are related to one another by either a sign inversion or a time shift, depending on whether BPSK or PPM was employed. Equations describing $s_0(t)$ and $s_1(t)$ can be obtained from Equations (3.83) and (3.84) by simply replacing the original pulse $p(t)$ with a rectangular pulse $\text{rect}(t)$ and setting b_j to $+1$ and -1, respectively. By using such template signals we are essentially performing a form of windowing. In other words, we are correlating the received signal against its underlying PN code over narrow windows.

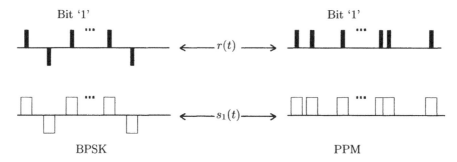

Fig. 3.15 The structure of the received and template signals.

The digital implementation of such a receiver entails sampling the received signal (once it has been sufficiently amplified), converting it to digital form, and then performing correlation. A figure analogous to Figure 3.16 describing this process can be obtained simply be replacing $r(t)$ and $s(t)$ with their discrete time counterparts $r(n)$ and $s(n)$, respectively.

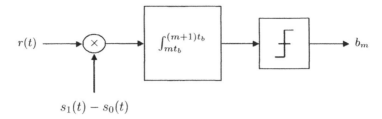

$$s_1(t) - s_0(t)$$

Fig. 3.16 Operations necessary for demodulation of a UWB signal.

3.5 DETECTION AND AMPLIFICATION

There are a number of approaches to the detection and amplification of the trains of UWB signals. In many cases there is an allocation of one-to-many in the assignment of bits to pulses to be transmitted. After 1945 the use of the correlation detection receiver became commonplace. Skolnik in his introductory book [28] written in 1962 describes the use of correlation methods in the detection of weak signals and cites a number of earlier references. A synchronous detector is also shown by Fink and Christiansen in [29]. There are a variety of ways to trigger the receiver on pulse rise time, level detection, integration over time, etc.

In the case of the correlation receiver detector, UWB and gate pulses are multiplied to produce a short-output, unamplified pulse whenever there is coincidence. Next, the result is fed to a (short-term) integrator or averager to produce a reduced amplitude, stretched signal output. If the integrator time is sufficiently long (conventional correlator) or a second long-term integrator is employed, the output will then represent the average of the many high-repetition-rate pulses fed to the correlator. Unfortunately, the integrator not only acts as a detector but also reduces the input amplitude in the step from narrow-pulse, low-duty cycle to averaged output. The long-term integrating correlator thus effects a many-to-one detection of averaged inputs prior to any amplification.

Micro-power impulse radar (MIR), or *radar on a chip*, offered an alternative to correlation detection [30]. The MIR is an integrating peak detector as opposed to the multiply-and-average correlation receiver detector described above. In the case of the MIR receiver detector, UWB and gate pulses are summed algebraically to form the input to a peak detector, i.e. the low-amplitude UWB pulse and the high-amplitude gate pulse, when summed, are above threshold for peak detection but individually are not. Moreover, it is not a single UWB pulse which, together with the gate pulse, provides the peak detected signal, but the (long-term) summing of a series of UWB inputs. The detector is thus triggered by the simultaneous occurrence of a summed series of low-amplitude UWB signals and a coincident large-amplitude gate pulse which, together, are algebraically summed. The coincident summing method of a large gate input and summed low-amplitude signals effects a many-to-one, peak signal detection process.

3.6 SUMMARY

In this chapter the basic tools for processing UWB signals and systems were introduced. Initially, the fundamentals of impulse electromagnetics in a lossy medium were
briefly explained. Basic tools for UWB waveform analysis in the time domain were
presented. The concepts of phase and instantaneous frequency as a measure of the
waveform informative value were introduced. Frequency response methods, such as
Fourier, Laplace, and z-transforms, were considered.

It should be emphasized that a specific UWB signal-processing technique does not
exist. Both time and frequency domain approaches can be applicable when analyzing
transmitter signaling, channel modeling, and antenna effect consideration. The most
significant difference between the UWB and narrowband systems is the variation of
major parameters with frequency. Due to this fact, working with UWB signals and
systems usually demands more complexity and higher degrees of calculations.

Problems

Problem 1. An LTI system has the following impulse response:

$$h(t) = 2e^{-at}u(t) \tag{3.86}$$

Use convolution to find the response $y(t)$ to the following input:

$$x(t) = u(t) - u(t-4) \tag{3.87}$$

Sketch $y(t)$ for the case when $a = 1$.

Problem 2. For each of the signals sketched in Figures 3.17(a)–(c) find the Fourier
transform and plot the magnitude and phase of the resulting functions.

Problem 3. Compute the DTFT of the following signals and sketch $X(e^{j\Omega})$:

(a) $x(n) = [1/4, 1/4, 1/4, 1/4]$;

(b) $x(n) = [1, -2, 1]$;

(c) $x(n) = 2(3/4)^n u(n)$.

Problem 4. Find the response of the following systems to the inputs below. Sketch
the magnitude of each frequency response for $-\pi < \Omega < \pi$ and determine the type
of filter:

$$x(n) = 2 + 2\cos\left(\frac{n\pi}{4}\right) + \cos\left(\frac{2n\pi}{3} + \frac{\pi}{2}\right) \tag{3.88}$$

(a) $H(e^{j\Omega}) = e^{-j\Omega}\cos(\Omega/2)$;

(b) $H(e^{j\Omega}) = e^{-j\Omega/2}(1 - \cos(\Omega/2))$.

Problem 5. Compute the Laplace transforms of the following functions:

(a) $x(t) = 4\sin(100t - 10)u(t - 0.1)$;

(b) $x(t) = tu(t) + 2(t - 2)u(t - 2) + (t - 3)u(t - 3)$;

(c) $x(t) = u(t) - e^{-2t}\cos(10t)u(t)$.

Problem 6. Sketch the response of each of the systems below to a step input:

(a) $H(s) = 10/s + 2$;

(b) $H(s) = 0.2/(s + 0.2)$;

(c) $H(s) = 1/(s^2 + 4s + 16)$.

Problem 7. Find the transfer functions of the following discrete time systems:

(a) $y(n) + 0.5y(n - 1) = 2x(n)$;

(b) $y(n) + 2y(n - 1) - y(n - 2) = 2x(n) - x(n - 1) + 2x(n - 2)$;

(c) $y(n) = x(n) - 2x(n - 1) + x(n - 2)$.

Problem 8. Given the following system,

$$y(n) = x(n) - x(n - 1) + x(n - 2) \tag{3.89}$$

(a) find the transfer function;

(b) give the impulse response;

(c) determine the stability;

(d) sketch the frequency response and determine the type of filter.

Problem 9. Solve the following difference equation using the z-transform:

$$y(n) + 3y(n - 1) + 2y(n - 2) = 2x(n) - x(n - 1) \tag{3.90}$$
$$y(-1) = 0$$
$$y(-2) = 1$$
$$x(n) = u(n)$$

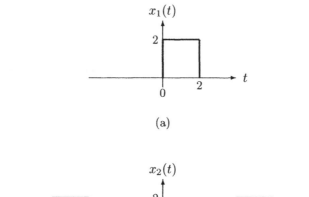

(a)

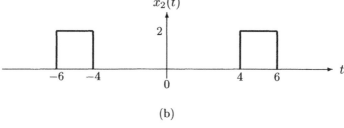

(b)

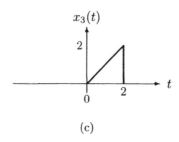

(c)

Fig. 3.17 Pulses for Problem 2.

4

UWB channel modeling

The propagation environment that a signal passes through from a transmitter to a receiver is referred to as the *channel*. Mobile cellular communication theory provides ample theory and measurement techniques for modeling signal propagation.

Thanks to its potential applications and unique capabilities, the UWB wireless communication system has been the subject of extensive research in recent years. However, many important aspects of UWB-based communication systems have not yet been thoroughly investigated. In particular, indoor and outdoor channel modeling and propagation effects require careful examination before an actual implementation of UWB systems can be undertaken. Without this, system performance and interference to and from other spectrum users cannot be readily ascertained.

The propagation of UWB signals in indoor and indoor–outdoor environments is the single most important issue, with significant impacts on the future direction, scope, and, generally, the extent of the success of UWB technology. If the channel is well characterized, the effect of disturbances and other sources of perturbation can be reduced by proper design of the transmitter and receiver. Detailed characterization of UWB radio propagation is, therefore, a major requirement for successful design of UWB communication systems.

One important aspect of any radio channel-modeling activity is the investigation of the distribution functions of channel parameters. Typically, these distributions are obtained from measurements or simulations based on almost exact or simplified descriptions of the environment. However, such methods often only yield insights into the statistical behavior of the channel and are not able to give a physical explanation of observed channel characteristics. Since these characteristics seem to yield complicated functions, both of the properties of wave propagation and of the individual geometry of the surroundings for which the channel is modeled, analytical models are rare.

Ultra Wideband Signals and Systems in Communication Engineering   Second Edition
M. Ghavami, L. B. Michael and R. Kohno © 2007 John Wiley & Sons, Ltd

The typical UWB propagation channel is a function which depends only weakly on the geometry of the environment. Rough knowledge about the surroundings is supposed to be sufficient for its characterization. Otherwise, no measurement campaign conducted in one environment could be a valid approximation of the channel in another, similar situation. Compared with the deterministic one, the stochastic channel should hence be analytically more tractable. However, in a suitable approach the robustness of the channel should result from the modeling process and not from an assumption on which the derivation is based.

In UWB radio channel modeling a number of aspects have to be taken into account that amount to a changed overall behavior of the channel. The main difference between UWB and traditional channel-modeling techniques is due to the fact that, in UWB propagation, frequency-dependent effects cannot be ignored. Parameters related to penetration, reflection, path loss, and many other effects should be considered frequency-variant and investigated more carefully.

4.1 A SIMPLIFIED UWB MULTIPATH CHANNEL MODEL

As a first example of channel-modeling techniques we examine a model which is not unique to UWB systems, but is considered a general model and is appropriate to explain the basic principles of channel modeling.

A convenient and simple model for characterization of UWB channels is the discrete time, multipath, impulse response model because the locations of ceilings, walls, doors, furniture, and people inside an indoor environment result in the transmitted signal taking multiple paths from the transmitter to the receiver (Figure 4.1). Hence, signals arrive at the receiver with different amplitudes, phases, and delays. In this model the delay axis is divided into small time intervals called *bins*. Each bin is assumed to contain either one multipath component or no component. The possibility of more than one path in a bin is excluded. A reasonable bin width is the resolution of a specific measurement, since two paths arriving within a bin cannot be resolved as distinct paths. Using this model, each impulse response can be described by a sequence of zeros and nonzeros, where a nonzero indicates the presence of a path in a given bin and a zero represents the absence of a path in that bin.

The phenomena of multipath propagation can be represented conveniently and mathematically by the following discrete impulse response of the channel:

$$h(t) = \sum_{l=0}^{L-1} \alpha_l \delta(t - lT_m) \qquad (4.1)$$

where α_l is the amplitude attenuation factor on path l and is a function of time and distance between the transmitter and receiver. The parameter T_m is the minimum resolution time of the pulse, L is the number of resolvable multipath components, and $\delta(t)$ is the Dirac delta function. Sometimes, Equation (4.1) is referred to as the *multipath intensity profile*.

For simplicity, in order to avoid partial correlations of the waveform, let the minimum path resolution time be equal to the modulated symbol period for which the

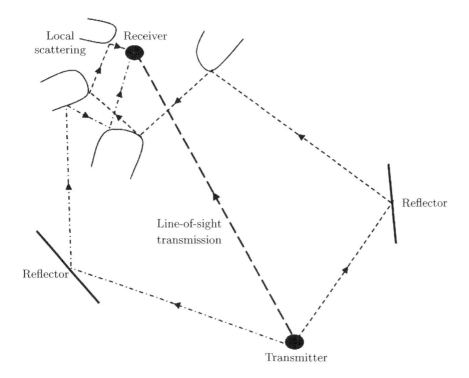

Fig. 4.1 A simple model of the indoor UWB radio multipath channel.

modulated waveform is nonzero (i.e. the UWB waveform is zero outside the interval $0 \leq t \leq T_m$).

According to Foerster [31] and Hashemi [32] the primary parameters that are important in the characterization of the indoor channel include the following:

- number of resolvable multipath components;

- multipath delay spread;

- multipath intensity profile;

- multipath amplitude-fading distribution;

- multipath arrival times.

The following subsections describe each of these components in more detail.

This simple multipath channel model and its parameters have been adopted, primarily based on those described by Hashemi [32], for several reasons. First, it is based upon a large number of measurements of various indoor channels and uses a minimum resolution time of 5 ns, which is representative of a broadband UWB channel. Second, the measurements are based on antenna separation distances of 5–30 m, which is typical of UWB-based systems, due to the expected PSD restrictions imposed by the FCC. Third, the analytical model described is relatively straightforward to realize for theoretical analysis or simulation.

4.1.1 Number of resolvable multipath components

The number of resolvable multipath components is important since it determines the design of a rake receiver. The distribution of the number of resolvable multipath components, L, for a large number of profiles collected in each building was investigated by Hashemi [32]. For each profile, L has been found by counting all multipath components within α dB of the strongest path, for example $\alpha = 10, 20, 30$ dB. The mean and standard deviation of L for each building, each value of α, and each transmitter–receiver antenna separation was collected. Examination of the data showed that:

(i) There is a clear dependence between the mean value of L and antenna separation.

(ii) The mean value of L increases with increasing α. This is expected since, when α increases, more components are included in the calculation of L.

(iii) The standard deviation of L increases with the increase in antenna separation. This is due to the fact that there are greater variations in the environment between the transmitter and receiver for large antenna separations. Also, standard deviations are very dependent upon the complexity and variations of the environment.

4.1.2 Multipath delay spread

Delayed signals through a channel are frequently described using one of the following three kinds of definitions of delay:

- *average delay* which describes mean travel time of a signal from a transmitter to a receiver;

- *delay spread* which is a metric of how much that signal is diluted in time;

- *maximum* or *total delay spread* which indicates the largest delay due to the multipath.

The delay spread of a UWB multipath channel is usually described by its rms value. Figure 4.2 demonstrates the delay profile, which is the expected power per unit of time received with a certain excess delay and is obtained by averaging a large set of impulse responses. The maximum delay time spread is the total time interval during which reflections with significant energy arrive. The rms delay spread is the standard deviation value of the delay of reflections, weighted proportional to the energy in the reflected waves. For a digital signal with high bit rate this dispersion is experienced as frequency-selective fading and ISI. No serious ISI is likely to occur if symbol duration is longer than, say, ten times the rms delay spread.

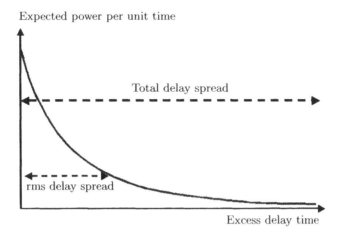

Fig. 4.2 A typical exponential delay profile with total and rms delay spread.

Typical values for the rms delay spread for indoor channels have been reported to be between 19 ns and 47 ns in [33], and mean values between 20 ns and 30 ns for 5 to 30 m antenna separation were reported in [32]. In addition, the multipath delay spread increases with increasing separation distance between receiving and transmitting antennas.

There also exists another relevant parameter, called the *Doppler spread*, that tells us how signal energy smears out in frequency in situations where the environment or the transmit/receive antennas move. Doppler spread can be important when the bandwidth of the UWB signal is very large or when the mobile has considerable movement.

4.1.3 Multipath intensity profile

The parameter rms delay spread T_{rms} is related to the multipath intensity profile in the sense that it represents the standard deviation of it. Therefore, if the form of the multipath intensity profile is known, then there should be a one-to-one relationship between rms delay spread and the multipath intensity profile. Results derived in [32] and many other references suggest that an exponentially (linear in dB) decaying multipath intensity profile is a reasonable model. This means that the average received power for path l can be represented by

$$E[\alpha_l^2] = \Omega_0 e^{-\delta l} \tag{4.2}$$

where Ω_0 is used to normalize the total received power to unity and δ is the decay factor.

The rms delay spread of the channel is utilized to determine the proper values for the two parameters, the total number of paths L and the decay factor δ. Since T_{rms} is simply the standard deviation of the multipath intensity profile, it can be determined in closed form as a function of L and δ. On the other hand, for a given value of T_{rms} there are an infinite number of (L, δ) pairs. In order to come up with a reasonable pair to estimate a channel model for a given T_{rms}, the following method can be used.

Initially, it should be considered that we are primarily interested in multipath components that have a power within 30 dB (0.001) down from the direct path (sometimes called the line-of-sight (LOS) path) component, since it is expected that subsequent paths will have a negligible effect on performance. As a result, we can write

$$e^{-L\delta} < 0.001 \tag{4.3}$$

Hence, for a given value for L we have

$$\delta \approx -\ln(0.001)/L \tag{4.4}$$

Now, we can find the smallest L that results in an actual T_{rms} greater than or equal to the desired T_{rms}, to represent a channel with the greatest decay factor.

4.1.4 Multipath amplitude-fading distribution

Based on the fading statistics described in [32], amplitude fades are best modeled by a log–normal distribution with a standard deviation between 3 dB and 5 dB for local distributions. In the log–normal distribution the logarithm of the random variable has a normal distribution. The probability density and cumulative distribution functions for the log–normal distribution are

$$P(x) = \frac{1}{Sx\sqrt{2\pi}} e^{-(\ln x - M)^2/(2S^2)} \tag{4.5}$$

and

$$D(x) = \frac{1}{2}\left[1 + \text{erf}\left(\frac{\ln x - M}{S\sqrt{2}}\right)\right] \tag{4.6}$$

respectively, where M and S determine the statistics of the log–normal distribution. The error function erf is defined as

$$\text{erf}(z) = \frac{2}{\sqrt{\pi}} \int_0^z e^{-t^2} dt \tag{4.7}$$

Note that the channel impulse response measurements described by Hashemi [32] have a resolvability of 5 ns. Based upon the intuitive understanding that UWB signals will probably experience less fading due to shorter pulse periods, the following results use a 3 dB standard deviation for the log–normal fading. Measurement results presented by Win and Scholtz [34] suggest that the fading is typically less than 5 dB for UWB impulse waveforms, which supports the smaller value for standard deviation. The mean value of log–normal fading is scaled such that it meets the exponentially decaying multipath intensity profile for the given delay spread of the channel.

4.1.5   Multipath arrival times

A simple statistical model for the arrival times of the paths is a Poisson process, since multipath propagation is caused by randomly located objects. Saleh and Valenzuela [35] have compared the path arrival distribution governed by a Poisson hypothesis with the empirical (measured) data to find the degree of closeness. The number of paths l in the first N bins of each measured profile was determined. To determine the empirical path index distribution the probability of receiving l paths in the first N bins $P_N(L = l)$ is plotted as a function of l. This procedure was repeated for different values of the number of bins N.

The probability $P_N(L = l)$ for the theoretical Poisson path index distribution is given by

$$P_N(L = l) = \frac{\mu^l}{l!} \qquad (4.8)$$

where l is the path index and μ is the mean path arrival rate given by

$$\mu = \sum_{i=1}^{N} r_i \qquad (4.9)$$

where r_i is the path occurrence probability for bin i computed from the empirical data.

Comparison of Poisson and empirical distributions in several papers, such as Suzuki [36] and Ganesh and Pahlavan [37], have revealed that there exists a considerable discrepancy between the two. Another model, called the modified Poisson model, was proposed by Suzuki [36].

The modified Poisson model is a simple and reasonable model for calculating the multipath arrival time of UWB systems. This model is also called the Δ-K model and has two states: state 1 and state 2. Whenever there is an event (i.e. a path occurs) the mean arrival rate, which is the average number of paths arriving per unit time or distance interval, is increased (or decreased if necessary) by a factor K for the next Δ seconds, where K and Δ are parameters to be chosen. The concept of a continuous Poisson model is illustrated in Figure 4.3.

Note that when $K = 1$ or $\Delta = 0$ the process returns to a standard Poisson sequence. For $K > 1$ the probability that there will be another path within the next Δ seconds increases (i.e. the process has a clustering property). For $K < 1$ the incidence of a path decreases the probability of having another path within the next Δ seconds (i.e. events have a tendency to arrive rather more equally spaced than in a pure Poisson model).

The analysis of this statistical model on the discrete delay time axis is rather simple and straightforward. It can be stated as the branching process of Figure 4.4, where λ_i ($i = 1, 2, \ldots$) is the underlying path occupancy rate for the ith bin and, for simplicity, Δ is taken as one 30 m time interval (empirical path occupancy rates are obtained for each 100 ft time interval) and K as a constant. The solid lines are to be traced when there was a path in the previous bin. The process starts at state 1 since

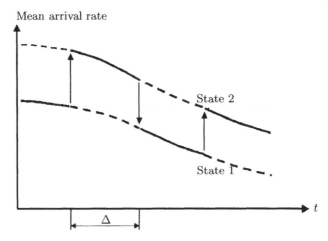

Fig. 4.3 Illustration of the modified Poisson process in the continuous case.

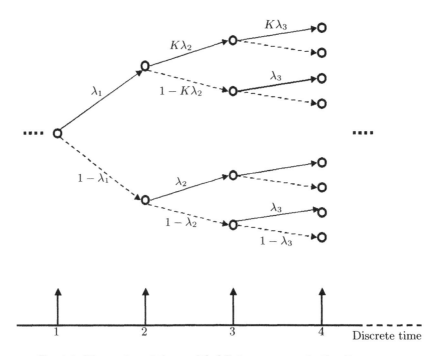

Fig. 4.4 Illustration of the modified Poisson process in the discrete case.

the LOS path is not counted as a path. Therefore, the probability to have a path in the first bin is λ_1. If a path does not occur in bin 1, then the probability of a path in bin 2 is λ_2, and if a path does occur in bin 1, then the probability of a path in bin 2 is $K\lambda_2$. The process proceeds similarly after this. For example, the probability that the process takes the arrowed route up to $N = 3$ is $\lambda_1(1 - K\lambda_2)\lambda_3$.

The path number distributions for this statistical model can be obtained, given the real (empirical) path occupancy rate r_i, by the following procedure. First, the underlying path occupancy rate λ_i is calculated from the values of r_i. As we easily see with the help of Figure 4.4, there exists the following relation between r_i and λ_i:

$$\lambda_i = \frac{r_i}{(K-1)r_{i-1}+1} \quad \text{for } i \neq 1 \tag{4.10}$$

where $\lambda_1 = r_1$.

4.2 PATH LOSS MODEL

By definition, the attenuation undergone by an electromagnetic wave in transit between a transmitter and a receiver in a communication system is called *path loss* or *path attenuation*. Path loss may be due to many effects, such as:

- free space loss;

- refraction;

- reflection;

- diffraction;

- clutter;

- aperture–medium coupling loss;

- absorption.

Note that path loss is usually expressed in decibels (dB). These effects are explained briefly in the following subsections.

4.2.1 Free space loss

Normally, the major loss of energy is due to the spreading out of the wave front as it travels away from the transmitter. As the distance increases, the area of the wave front spreads out, much like the beam of a torch. This means the amount of energy contained within any unit of area on the wave front will decrease as distance increases. By the time the energy arrives at the receiving antenna the wave front is so spread out that the receiving antenna extends into only a very small fraction of the wave front.

Free space loss is the signal attenuation that would result if all absorbing, diffracting, obstructing, refracting, scattering, and reflecting influences were sufficiently removed so as to have no effect on propagation.

4.2.2   Refraction

Refraction is defined, in the general case, as redirection of a wave front passing through (a) a boundary between two dissimilar media or (b) a medium having a refractive index that is a continuous function of position (e.g., a graded index optical fiber). For two media of different refractive indices the angle of refraction is closely approximated by *Snell's law*.

4.2.2.1   Snell's law   Snell's law is originally a law of geometric optics that defines the amount of bending that takes place when a light ray strikes a refractive boundary (e.g., an air–glass interface) at a nonnormal angle. Snell's law states that

$$n_1 \sin \theta_1 = n_2 \sin \theta_2 \tag{4.11}$$

where n_1 is the index of refraction of the medium in which the incident ray travels; θ_1 is the incident angle, with respect to the normal at the refractive boundary, at which the incident ray strikes the boundary; n_2 is the index of refraction of the medium in which the refracted ray travels; and θ_2 is the angle, with respect to the normal at the refractive boundary, at which the refracted ray travels. The incident ray and refracted ray travel in the same plane, on opposite sides of the normal at the point of incidence.

If a ray travels from a medium of lower refractive index into a medium of higher refractive index, it is bent toward the normal; if it travels from a medium of higher refractive index to a medium of lower index, it is bent away from the normal. If the incident ray travels in a medium of higher refractive index toward a medium of lower refractive index at such an angle that Snell's law would call for the sine of the refracted ray to be greater than unity (a mathematical impossibility), i.e.

$$\sin \theta_2 = \frac{n_1}{n_2} \sin \theta_1 > 1 \tag{4.12}$$

then the *refracted* ray in actuality becomes a *reflected* ray and is totally reflected back into the medium of higher refractive index, at an angle equal to the incident angle (and thus still obeys Snell's law). This reflection occurs even in the absence of a metallic reflective coating (e.g., aluminum or silver). This phenomenon is called *total internal reflection*. The smallest angle of incidence, with respect to the normal at the refractive boundary, which supports total internal reflection, is called the *critical angle*.

4.2.3   Reflection

The abrupt change in direction of a wave front at an interface between two dissimilar media so that the wave front returns into the medium from which it originated is called *reflection*. Reflection may be *specular* (mirror-like) or *diffuse* (i.e. not retaining the image, only the energy) according to the nature of the interface. Depending upon the nature of the interface (i.e. dielectric–conductor or dielectric–dielectric) the phase of the reflected wave may or may not be inverted.

4.2.4 Diffraction

Diffraction is the *spreading out* of waves. All waves tend to spread out at the edges when they pass through a narrow gap or pass an object. Instead of saying that the wave spreads out or bends round a corner, we can say that it diffracts around the corner.

A narrow gap is one which is about the same size as the wavelength of the electromagnetic wave or less. The longer the wavelength of a wave, the more it will diffract.

4.2.5 Wave clutter

Disorganized wave propagation due to a rough surface or interface is called wave clutter. The mechanisms leading to clutter are not well known so far, and they are much more complex compared with the narrowband case. Understanding the phenomenology of electromagnetic interactions between very short pulses and the complex dielectric ground surface provides an important input to the design of UWB systems, leading to improved clutter cancellation and improved detection performance. The roughness considerably influences the spectral content of the response.

4.2.6 Aperture–medium coupling loss

Coupling loss is the loss that occurs when energy is transferred from one medium to another. Aperture–medium coupling loss is the difference between the theoretical gain of the antenna and the gain that can be realized in operation. Aperture–medium coupling loss is related to the ratio of *scatter angle* to *antenna beamwidth*.

4.2.7 Absorption

In the transmission of electrical, electromagnetic, or acoustic signals the conversion of the transmitted energy into another form, usually thermal, is called *absorption*. Absorption is one cause of signal attenuation. The conversion takes place as a result of interaction between the incident energy and the material medium, at the molecular or atomic level.

4.2.8 Example of free space path loss model

There have been several proposed path loss models in the literature (e.g., see Cramer *et al.* [38] and Ghassemzadeh *et al.* [39]). Assuming perfect isotropically radiating antennas at the transmitter and receiver, the received power as a function of frequency can be expressed as [40]

$$P_R(f) = \frac{P_T(f)G_T(f)G_R(f)c^2}{(4\pi d)^2 f^2} \qquad (4.13)$$

where $P_T(f)$ is the average transmit power spectral density, c is the speed of light, and $G_T(f)$ and $G_R(f)$ are the transmit and receive antenna frequency responses, respectively.

Clearly, Equation (4.13) depends on the frequency response of the antennas, which may be difficult to generalize, especially when we are dealing with wideband signals. However, the present regulations for UWB requires the transmitter to meet a certain electric field strength limit at a specified range, which is equivalent to a total, limited transmit PSD.

It is desirable that the product $P_T(f)G_T(f)$ be flat within the bandwidth of interest. Hence, as a first-order approximation, a flat frequency response isotropic antenna is considered. Therefore, for a perfectly flat UWB waveform occupying the band $f_c - W/2$ to $f_c + W/2$ with PSD P_{av}/W and a flat frequency response of the receiving antenna with constant gain across the whole bandwidth (G_R), the total average received power at the output of the receiving antenna will be given by the following equation:

$$
\begin{aligned}
P_{Rav} &= \int_{f_c - W/2}^{f_c + W/2} P_R(f)\, df \\
&= \frac{P_{av} G_R c^2}{W(4\pi d)^2} \left[\frac{1}{f_c - W/2} - \frac{1}{f_c + W/2} \right] \\
&= \frac{P_{av} G_R c^2}{W(4\pi d)^2 f_c^2} \left[\frac{1}{1 - (W/2f_c)^2} \right]
\end{aligned}
\tag{4.14}
$$

which can be equally written as

$$
P_{Rav} = P_{av}^{NB} \left[\frac{1}{1 - (W/2f_c)^2} \right]
\tag{4.15}
$$

where

$$
P_{av}^{NB} = \frac{P_{av} G_R c^2}{W(4\pi d)^2 f_c^2}
\tag{4.16}
$$

corresponds to the well-known, narrowband, free space, path loss model equation. Hence, the second term indicated in Equation (4.15) accounts for the difference between the narrowband and wideband models. For the largest fractional bandwidth allowed by current UWB regulations (occupying 3.1–10.6 GHz), P_{Rav} will differ from P_{av}^{NB} by only 1.5 dB, and this difference becomes smaller for smaller fractional bandwidths. Also note that the FCC rules result in $W < 2fc$, so the singularity in the above equation can be ignored at $W = 2fc$.

Alternatively, we can repeat the above analysis for a receiver antenna response of the following form:

$$
G_R(f) = \frac{4\pi A_R f^2}{c^2}
\tag{4.17}
$$

where A_R is the effective area of the antenna (e.g., the antenna has a fixed effective aperture). This type of response yields a greater gain for higher frequencies. In this case the above analysis results in the average received power given by

$$P_{Rav} = \frac{P_{av}4\pi A_R}{(4\pi d)^2}$$

$$= \left[\frac{P_{av}c^2}{(4\pi d)^2 f_c^2}\right]\left[\frac{4\pi A_R f_c^2}{c^2}\right]$$

$$= P_{av}^{NB}G_R(f_c) \tag{4.18}$$

where $G_R(f_c)$ is the antenna gain at the center frequency of the transmitted waveform.

In conclusion, it appears that the narrowband model can be used to approximate the path loss for a UWB system, based on the assumptions discussed above.

4.3 TWO-RAY UWB PROPAGATION MODEL

Path loss is considered to be a fundamental parameter in channel modeling as it plays a key role in link budget analysis. Also, path loss serves as an input for the mean value of large-scale fading, which in turn determines the small-scale fading characteristics. In many channel models, path loss is modeled by adopting a power-law dependence with distance from the transmitter $L_p = ad^\gamma$, which is empirically extracted from data collected by measurement campaigns. In some cases a simple free space law, where $\gamma = 2$, with additional losses is adopted [41]. Otherwise, the values of the exponent γ and the coefficient a of the mean path loss are either obtained through linear regression to fit empirical data [42] or statistically characterized by means of probability distributions [39]. A random variable with log–normal distribution is usually added to the mean path loss that takes into account variations associated with the shadowing phenomenon, in order to model the fading.

Path loss frequency dependence has sometimes been excluded in UWB propagation channel modeling, and, although its impact is negligible over the bandwidths of current wireless systems, the same assumption cannot be applied to UWB systems. This section describes a theoretical study of path loss over short ranges. A rigorous formulation of a two-ray link is considered and evaluated as a function of both frequency and distance, where all the assumptions adopted for deriving the traditional narrowband plane earth model are no longer valid. Considering the specific UWB application the analysis is carried out over distances of up to 10 m and in the operational frequency band of 3 to 10 GHz.

From this investigation an analytical path loss model for short-range, two-ray links with wide operational bandwidths is derived, which incorporates the frequency dependence; however, it does not consider the effect of the transmitter and receiver antennas. This simple model is applicable to LOS UWB transmissions. The effect of path loss frequency selectivity and its impact on UWB pulse shape are also demonstrated. The analysis is performed by modeling the path loss as a low-pass filter and applying the associated impulse response to the UWB pulse signal. The observations confirm that path loss frequency dependence cannot be neglected in UWB systems, due to the extremely large bandwidth of the involved signals. Thus, the following issues [43] are emphasized in this section:

- the path loss frequency dependence;

- the observation of a breakpoint at short distances of about 3 m;

- the filtering effect of path loss on UWB transmissions.

4.3.1 Two-ray path loss

A two-ray channel model is widely considered to be a good approximation for LOS UWB links operating in a relatively clutter-free environment. According to the classical definition of two-ray propagation, the direct ray and the ray reflected by the ground are considered, as illustrated in Figure 4.5.

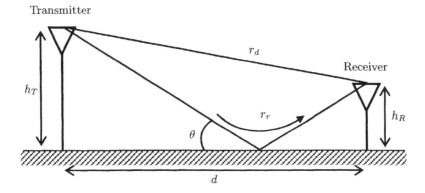

Fig. 4.5 Geometry of the two-ray model including a transmitter and a receiver.

The ground is regarded as a plane surface at this stage; thus, specular reflection is assumed. Indeed, the degree of roughness of a surface is frequency dependent and results in an angular spread of the reflected field. Its effect on UWB radio transmissions is a topic of research and needs to be incorporated into subsequent models.

Considering the field at the receiver as the superposition of the contributions associated with the two rays, the path loss can be expressed as $L_p = (G_p)^{-1}$, where the path gain G_p is defined as

$$G_p = \left(\frac{\lambda}{4\pi}\right)^2 \left| \frac{e^{-jkr_d}}{r_d} + \frac{R_{H,V}e^{-jkr_r}}{r_r} \right|^2 \tag{4.19}$$

and $k = 2\pi/\lambda$ is the free space propagation constant, $\lambda = c/f$ is the free space wavelength, and c is the speed of light. Moreover, as shown in Figure 4.5,

$$r_d = \sqrt{d^2 + (h_T - h_R)^2} = d\sqrt{1 + \left(\frac{h_T - h_R}{d}\right)^2} \qquad (4.20)$$

and

$$r_r = \sqrt{d^2 + (h_T + h_R)^2} = d\sqrt{1 + \left(\frac{h_T + h_R}{d}\right)^2} \qquad (4.21)$$

are the lengths of the direct path and of the reflected path, respectively. In Equation (4.19), $R_{H,V}$ indicates the Fresnel reflection coefficient for the horizontal or vertical polarization, defined as

$$R_H = \frac{\sin\theta - \sqrt{\varepsilon_r - \cos^2\theta}}{\sin\theta + \sqrt{\varepsilon_r - \cos^2\theta}} \qquad (4.22)$$

and

$$R_V = \frac{\varepsilon_r \sin\theta - \sqrt{\varepsilon_r - \cos^2\theta}}{\varepsilon_r \sin\theta + \sqrt{\varepsilon_r - \cos^2\theta}} \qquad (4.23)$$

where

$$\varepsilon_r(f) = \varepsilon_r'(f) - j\frac{\sigma(f)}{2\pi f \varepsilon_0} \qquad (4.24)$$

is the dielectric constant of the reflecting surface. The frequency dependence of the relative permittivity $\varepsilon_r'(f)$ and the conductivity $\sigma(f)$ must be taken into account to appropriately characterize reflection and transmission mechanisms over the UWB frequency range.

In narrowband outdoor and indoor modeling, construction material properties have been defined for a fixed operating frequency. A database, obtained from power loss measurements in the frequency range of 3 to 8 GHz, has been presented by Stone [44] where power attenuation through a set of construction materials with different thicknesses is described as a function of frequency. However, numerical estimates on dielectric constant and permittivity have not been provided as yet. Hence, at the present stage we consider constant values of $\varepsilon_r'(f) = \varepsilon_r'$ and $\sigma(f) = \sigma$.

The traditional plane earth model, as described by Rappaport [45], which is used in classical narrowband transmissions at long distances, and even in microcells and picocells, is *not* applicable to short ranges, since the two hypotheses on which its derivation is based, namely that the distance between transmitter and receiver is much greater than the antenna heights $d \gg h_T$ and $d \gg 4h_T h_R/\lambda$, are not satisfied. This implies that all the approximations held in the classical two-ray model [45] (e.g., reflection angle $\theta \cong 0$ and reflection coefficient $|R_{H,V}| \cong 1$) are no longer valid and, obviously, the well-known fourth-order power dependence on distance does not hold for distances below the breakpoint $4h_T h_R/\lambda$. Thus, a rigorous analytical evaluation is required for short ranges, when the distance d is comparable with h_T and h_R. By replacing Equations (4.20) and (4.21) in Equation (4.19), the path gain can be written as

$$G_p(d,f) = G_{FS}(d,f) \left| \frac{e^{-jkr_d}}{\sqrt{1+((h_T-h_R)/d)^2}} + \frac{R_{H,V}e^{-jkr_r}}{\sqrt{1+((h_T+h_R)/d)^2}} \right|^2 \qquad (4.25)$$

where

$$G_{FS}(d,f) = \left(\frac{c}{4\pi df} \right)^2 \qquad (4.26)$$

is the classical free space path gain depending on the transmitter–receiver distance d. Note that for short ranges the free space transmission distance r_d cannot in general be approximated with d, and as apparent in Equation (4.20), this approximation depends on the ratio $(h_T - h_R)/d$.

The path gain in Equation (4.25) is clearly a function of both distance and frequency, as shown in Figure 4.6. Here Equation (4.25) is evaluated in the case of vertical polarization of the field and total reflection from the ground ($\sigma \to \infty$, $R_V = 1$) and plotted in decibels, $L_{p_{dB}} = -10\log(G_p)$, versus both distance and frequency.

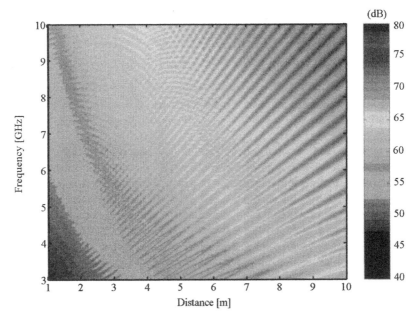

Fig. 4.6 Path loss versus distance and frequency: $h_T = 2.5$ m, $h_R = 1.2$ m, and $R_V = 1$.

The two-ray path loss clearly exhibits fading around a mean value due to the variation in the phase of the two-ray contributions. Therefore, mean path loss can be modeled as

$$\bar{L}_p(d, f) = \alpha d^{\gamma(f)} f^{\nu(d)} \tag{4.27}$$

where $d_{\min} \leq d \leq d_{\max}$ and $f_{\min} \leq f \leq f_{\max}$. The corresponding expression in decibels is written as

$$\bar{L}_{p_{\mathrm{dB}}}(d, f) = \alpha_{\mathrm{dB}} + \gamma(f)10\log(d) + \nu(d)10\log(f) \tag{4.28}$$

where $\gamma(f)$ and $\nu(d)$ turn out to be the slope coefficients and are functions of frequency and distance, respectively. In particular, at each frequency point f^*, the slope coefficient $\gamma(f^*)$ is determined by linear fitting of the decibel path loss curve, which results in a function of distance d only (i.e. $L_{p_{\mathrm{dB}}}(d)$ at $f = f^*$). Likewise, at each distance point d^* the value $\nu(d^*)$ is determined by linear fitting the curve $L_{p_{\mathrm{dB}}}(f)$ at $d = d^*$.

4.3.2   Two-ray path loss model

An analysis of Equation (4.28) as a function of distance and frequency enables a power-law path loss model to be derived:

$$\bar{L}_{p_{\mathrm{dB}}}(d, f) \Big|_{\substack{f_{\min} \leq f \leq f_{\max} \\ d_{\min} \leq d \leq d_{\max}}} = L_{p0} + \gamma 10\log\left(\frac{d}{d_0}\right)\Big|_{f_{\min} \leq f \leq f_{\max}}$$

$$+ \nu 10\log\left(\frac{f}{f_0}\right)\Big|_{d_{\min} \leq d \leq d_{\max}} \tag{4.29}$$

where the slope coefficients γ and ν are constant values and

$$L_{p0} = 10\log(L_{\mathrm{FS}}(r_{d_0}, f_0)) \tag{4.30}$$

is the free space path loss at $r_{d_0} = r_d(d_0)$, where d_0 is a reference distance, and f_0 is a reference frequency. The values of γ and ν are obtained by averaging $\gamma(f)$ and $\nu(d)$ in the ranges $f_{\min} \leq f \leq f_{\max}$ and $d_{\min} \leq d \leq d_{\max}$.

In particular, in the distance range d from 1 to 10 m and in the frequency range of f between 3 GHz and 10 GHz the path loss model exhibits the following features:

- **Distance dependence:** *The path loss has dual-slope behavior with respect to distance*, with a new breakpoint at a very short distance from the transmitter (a few meters). The exact position of the breakpoint d_{BP} occurs at a closer distance from the transmitter as the difference $\Delta h = h_T - h_R$ is reduced. The breakpoint defines two sub-ranges, $1\,\mathrm{m} \leq d_1 \leq d_{\mathrm{BP}}$ and $d_{\mathrm{BP}} \leq d \leq 10\,\mathrm{m}$, over which $\gamma(f)$ is evaluated. The values $\gamma = \gamma_l$ and $\gamma = \gamma_u$ for use in Equation (4.29) are obtained as the mean value of $\gamma(f)$. Note that this breakpoint is valid at short distances and for operation over an extremely wide frequency range. Moreover, it should be emphasized that it is different from the breakpoint of the traditional plane earth model $4h_T h_R/\lambda$, which is clearly frequency dependent and is located outside the maximum distance considered in this analysis and, therefore, outside of the expected operating range of UWB transmissions.

- **Frequency dependence:** *The path loss shows a $\nu = 2$ power-law frequency dependence*, which is the same as the frequency dependence of free space transmissions (i.e. only a single ray). This result is due to the fact that the frequency dependences of permittivity, conductivity, and surface roughness have not been taken into account. The frequency dependence of the material properties and variations of ν are expected, confirming the frequency dependence of the clutter impact over the UWB frequency range.

The model in Equation (4.29) is valid for both horizontal and vertical polarizations, as the two polarizations show the same mean path loss tendency. In the particular case of very conductive surfaces (i.e. the reflection coefficient's absolute value is close to 1) the oscillations around the mean value have the same amplitude for both polarizations, but are opposite in phase; averaging the two path loss curves gives a fade-free path loss curve that is in close agreement with the power-law path loss model.

Some results are shown in Figures 4.7–4.10, where the transmitter height is $h_T = 2.5$ m and the receiver height is $h_R = 1.2$ m. In this particular case the new short-distance breakpoint is located at $d_{\mathrm{BP}} = 3.5$ m from the transmitter. Note that, with this choice of h_T and h_R, the location of the traditional breakpoint $4h_T h_R/\lambda$, from which a fourth power dependence is assumed in the classical plane earth model, would range between 120 m at 3 GHz and 400 m at 10 GHz. A relative permittivity $\varepsilon_r' = 6$ and a conductivity $\sigma = 0.0166$ are assumed when computing the plane reflection coefficient Equation (4.24) and modeling the dielectric constant of a cement surface.

Figure 4.7 shows the plots of $\gamma(f)$ that result from evaluation over lower and upper distance sub-ranges separated by the breakpoint: 1 m $\leq d_l \leq 3.5$ m in Figure 4.7(a) and 3.5 m $\leq d_u \leq 10$ m in Figure 4.7(b). The values of γ are $\gamma_l = 1.32$ and $\gamma_u = 1.9$, respectively. Note that the values of the slope coefficient resemble those provided in previous literature [39], [42] for LOS UWB transmissions, but values of the slope coefficient have not been included when the breakpoint is at very short distance from the transmitter. Similarly, in Figure 4.8 the slope coefficient $\nu(d)$ is evaluated over the entire frequency range 3 GHz $\leq f \leq 10$ GHz and plotted versus distance; its mean value is $\nu = 2$.

In Figure 4.9, the path loss is plotted in decibels versus the distance d at the operational frequency $f = 5$ GHz. Note that this case does not show a UWB situation. The two curves exhibiting fading are obtained by evaluating the rigorous analytical formulation of the two-ray path loss, and each one is related to a polarization. The solid line corresponds to the path loss model, with the breakpoint at $d_{\mathrm{BP}} = 3.5$ m, and the dotted line is the free space path loss.

In Figure 4.10 the path loss evaluated at a distance $d = 2$ m is plotted versus frequency. Again, path loss curves for both horizontal and vertical polarizations are plotted: the solid line is the result obtained through our path loss model in Equation (4.29), with $\nu = 2$, and the dotted line, which is 3 dB below, is the free space path loss evaluated at the same distance as a variable of frequency.

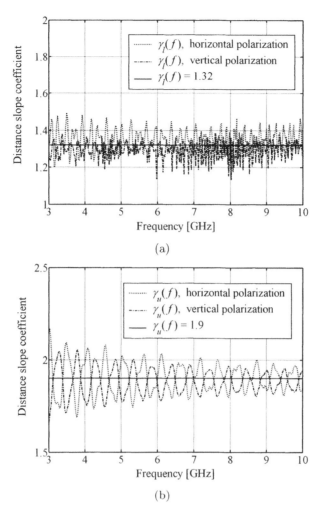

Fig. 4.7 Path loss distance slope coefficient $\gamma(f)$ and mean value γ, for 3 GHz $\leq f \leq$ 10 GHz, $h_T = 2.5$ m, $h_R = 1.2$ m: (a) 1 m $\leq d \leq 3.5$ m, $\gamma_l = 1.32$; (b) 3.5 m $\leq d \leq 10$ m, $\gamma_u = 1.9$.

4.3.3 Impact of path loss frequency selectivity on UWB transmission

UWB systems are clearly different from classical modulated or passband radio systems. Likewise, UWB channel modeling must consider the true nature of the UWB propagation channel. In particular, it is well known that the frequency selectivity of propagation mechanisms leads to distortion of signals arriving at the receiver. It is therefore necessary to investigate the impact on ISI and bit error rate (BER) performance of the system under such conditions. Provided the frequency response of the channel is known, it can be combined with the frequency spectrum of the UWB pulse and the distortion level of the received signal can be evaluated. The frequency

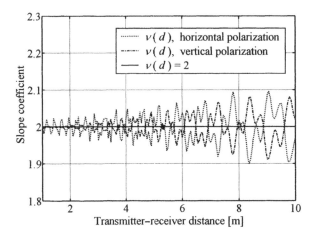

Fig. 4.8 Path loss frequency slope coefficient $\nu(d)$ and mean value $\nu = 2$ for $1 \text{ m} \leq d \leq 10 \text{ m}$, $h_T = 2.5$ m, and $h_R = 1.2$ m.

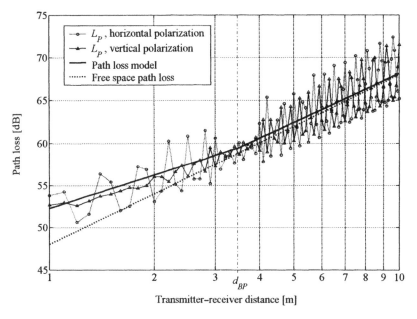

Fig. 4.9 Path loss versus distance for $1 \text{ m} \leq d \leq 10 \text{ m}$, $f = 5 \text{ GHz}$, $h_T = 2.5$ m, and $h_R = 1.2$ m.

selectivity of the path loss has been neglected in link-level investigations of wireless systems until now.

In the following, a $\nu = 2$ frequency power law (i.e. free space) is assumed for the path loss and is in line with the observations made before, which neglect antenna effects. This produces a low-pass filtering effect on the transmitted signal

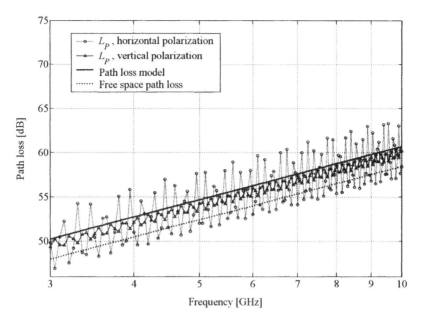

Fig. 4.10 Path loss versus frequency for 3 GHz $\leq f \leq$ 10 GHz, $d = 2$ m, $h_T = 2.5$ m, and $h_R = 1.2$ m.

(see Figure 4.10). In fact, it can be modeled as a transfer function in the frequency domain by considering the path gain as

$$G_p = |H_p(f)|^2 \qquad (4.31)$$

Hence, from Equation (4.19), when assuming no reflected ray, the following transfer function can be defined:

$$H_p(f) = K_H \frac{c}{2d} \frac{e^{-j(2\pi/c)\,df}}{2\pi f} \qquad (4.32)$$

where d is the distance between the transmitter and the receiver and K_H is the normalization constant necessary to meet the requirements on the energy associated with the filter.

By applying Fourier theory the corresponding path gain impulse response in the time domain is determined:

$$h_p(t) = j \frac{H_0}{2} \operatorname{sign}(t - t_d) \qquad (4.33)$$

where

$$H_0 = K_H \frac{c}{4d} \qquad (4.34)$$

and $t_d = 2\pi d/c$ is the propagation time delay between transmitter and receiver. The normalized impulse response $h_p(t)/(jH_0/2)$ is plotted in Figure 4.11 for a $d = 2$ m link that gives $t_d = 6.67$ ns.

By convoluting the path gain impulse response $h_p(t)$ with the transmitted signal $s(t)$, we obtain the signal at the receiver:

$$w(t) = s(t) * h_p(t)$$

$$= jH_0 \int s(t)\, dt \qquad (4.35)$$

Thus, we can conclude that free space propagation produces an integration effect on a signal occupying a wide frequency spectrum. As an example, consider a typical UWB signal pulse $s(t)$ that is the second derivative of a Gaussian waveform,

$$s(t) = K_s \frac{2}{\tau^2} \left(\frac{t^2}{\tau^2} - 1 \right) e^{-[t/\tau]^2} \qquad (4.36)$$

and apply $h_p(t)$; as a result, the signal collected at a receiver located at $d = 2$ m from the transmitter is

$$w(t) = -j2\pi H_0 K_s \frac{t - t_d}{\tau} e^{-[(t-t_d)/\tau]^2} \qquad (4.37)$$

Figures 4.11(b) and 4.11(c) show the waveforms defined in Equations (4.36) and (4.37), respectively, with $\tau = 0.5$ ns.

In conclusion, neglecting the frequency dependence of path loss in UWB propagation channel models seems to be an unsuitable choice, as the filtering effect on the UWB pulse shape is not considered, with possible consequences on the development of an effective UWB radio system. For example, the knowledge of the distortion to which the pulse is exposed would help in the choice of a suitable correlator when designing a UWB coherent receiver. Note, at this stage observation has been limited to the propagation channel (i.e. path loss) and antennas have not been included in the analysis. Nevertheless, effective modeling will also consider the pass-band behavior of transmitting and receiving antennas, which will contribute to the overall channel-filtering effect.

4.4   FREQUENCY DOMAIN AUTOREGRESSIVE MODEL

In this section an autoregressive (AR) frequency domain statistical model for UWB indoor radio propagation is described. Using AR modeling techniques the parameters of the channel model can be determined from the measured frequency responses. A frequency domain channel-sounding experiment is usually performed for this purpose.

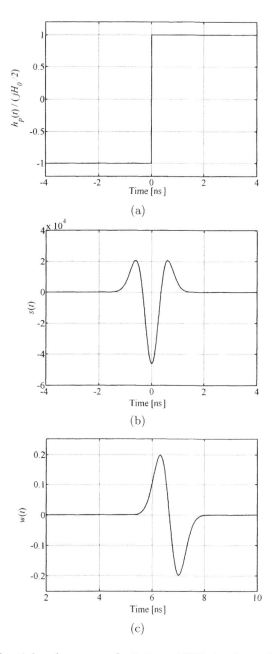

Fig. 4.11 Impact of path loss frequency selectivity on UWB signal waveforms: (a) normalized impulse response; (b) transmitted pulse waveform; (c) received pulse waveform.

By definition, an AR model depends only on the previous outputs of the system. The transfer function representation of such a model can be written as follows:

$$H(z) = \frac{Y(z)}{X(z)}$$

$$= \frac{b_0}{1 - a_1 z^{-1} - a_2 z^{-2} - \cdots} \qquad (4.38)$$

where $X(z)$ and $Y(z)$ denote the z-transforms of the input and output of the AR system, respectively. Equivalently, in the discrete time domain we can write Equation (4.38) as follows:

$$y(n) = b_0 x(n) + a_1 y(n-1) + a_2 y(n-2) + \cdots \qquad (4.39)$$

The target of a frequency domain UWB channel model is to develop a statistical representation of the channel with a minimum number of parameters to regenerate the measured channel behavior accurately in computer simulations. The higher the complexity of the model, the closer the statistical resemblance to that of measured data [46].

The main advantage of the frequency domain modeling of systems over time domain modeling is that it uses fewer parameters. On the other hand, it requires a complete characterization of the probability distributions of these parameters.

The frequency response of a system can be interpreted as the output of an AR model [47]. AR modeling of time domain data used for spectral estimation and the techniques for determination of the coefficients or multipliers of the AR process are well known in the literature [48]. These results can be used for AR modeling of the frequency response observed from propagation measurements.

With the AR process assumption the frequency response at each location is a realization of an AR process of order p given by the following equation:

$$H(f_n, x) - \sum_{i=1}^{p} a_i H(f_{n-1}, x) = V(f_n) \qquad (4.40)$$

where $H(f_n, x)$ is the nth sample of the complex frequency domain measurement at location x and $\{V(f_n)\}$ is a complex white noise process. The parameters of the model are the complex constants a_i. Taking the z-transform of Equation (4.40), we can view the AR process $\{H(f_n, x)\}$ as the output of a linear filter with transfer function

$$G(z) = \frac{1}{1 - \sum_{i=1}^{p} a_i z^{-i}}$$

$$= \prod_{i=1}^{p} \frac{1}{1 - p_i z^{-i}} \qquad (4.41)$$

driven by a zero-mean white Gaussian noise process $\{V(f_n)\}$. Using the AR model the channel frequency response can be identified with the p parameters of the AR model or the location of the p poles of the transfer function $G(z)$.

An AR model can be implemented using an infinite impulse response (IIR) filter as shown in Figure 4.12. In a simple AR model with a second-order IIR filter [46], the frequency response model has four complex parameters and one real parameter, which is the noise standard deviation σ. Thus, the model is characterized by nine real parameters. For a model at a particular location with a transmitter–receiver separation of 0.6 m the estimated parameters are given in [46].

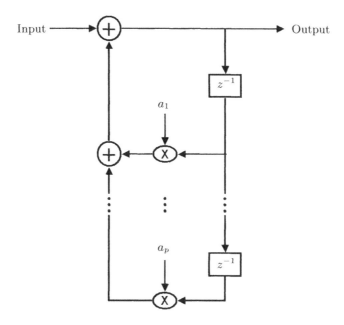

Fig. 4.12 Implementation of an AR model using an IIR formulation.

The agreement between the probability distribution of the model parameters using experimental data and the normal distribution is also shown for a particular location. Using this model the channel frequency response has been simulated as well. To verify model accuracy a comparison was made between the modeled probability density and that of the measured data. The model has been reported to reproduce the frequency selectivity and the multipath propagation characteristics observed in actual measurements.

4.4.1 Poles of the AR model

The AR model can be represented by the poles of its transfer function given by Equation (4.41),

$$1 - \sum_{i=1}^{p} a_i z^{-i} = \prod_{i=1}^{p}(1 - p_i z^{-i}) \tag{4.42}$$

It has been demonstrated that the largest pole is very close to the unit circle. A pole close to the unit circle represents significant power at the arrival time related to the angle of the pole. The arrival times of significant paths in all measurements are from 3 to 74 ns. The delay is calculated as

$$\tau = \frac{\arg(p_i)}{2\pi f_s} \tag{4.43}$$

where f_s is the sampling frequency. Experimenting with higher-order models, it was observed that two significant poles exist in the indoor radio propagation range and this is adequate to represent the in-home UWB channel. The two significant poles can be interpreted as two significant clusters of multipath arrivals. The interpretation of a pole as defining a cluster of paths and the distance of the pole from the unit circle as defining the power in the cluster provides a useful physical interpretation of the AR model [47]. In contrast to other wideband systems, the UWB channel represents itself as two pole clusters with a distinctive angular spread.

4.5   IEEE PROPOSALS FOR UWB CHANNEL MODELS

Due to the large bandwidth (3.6–10.1 GHz) allocated to UWB systems, a single-band UWB signal propagates differently in the lower and upper bands of the spectrum. The higher frequency components of the signal can be obstructed more easily and man-made radio interference is not homogenously distributed over the whole spectrum. The first experimental results were obtained by [49] for a UWB channel and they showed a short-time duration of the channel impulse response, a strong path loss attenuation, and a great number of resolvable multipaths.

Following this first effort, a lot of experimental measurements have been undertaken to characterize accurately the UWB channel, taking into account the ultra wide spectrum in different radio environments. The results of all these contributions were analyzed and compiled in a report [50] edited by the channel modeling subcommittee of the IEEE 802.15.3a working group for wireless personal area networks, in 2003.

Following the consensus reached in this report, a reference UWB channel model was released for indoor and high-data-rate UWB applications known as the IEEE 802.15.3a channel model. This report also includes a Matlab program to generate the impulse response of this model. In December 2004, another reference UWB channel model was released, this time for outdoor and low-data-rate UWB applications known as the IEEE 802.15.4a channel model [51].

The IEEE 802.15.3a channel model encompasses the main features of the Saleh–Valenzuela channel model [35], where multipath rays arrive in clusters and their amplitudes follow a double-exponential decay. Nevertheless, this novel channel model differs from the Saleh–Valenzuela model in terms of the multipath amplitude distribution, which has been reported in [50] to be log–normal. Thus, each multipath gain magnitude coefficient of this frequency-selective channel model is assumed to be a log–normally distributed random variable. This model is designed for baseband signalling, so therefore the phase of the channel impulse response is either 0 or π.

4.5.1   An analytical description of the IEEE UWB indoor channel model

The IEEE 802.15.3a channel model is based on the Saleh–Valenzuela model, where multipath rays arrive in clusters. The consecutive cluster times of arrival $T_c - T_{c-1}$ are well modeled by random variables which follow exponential probability distributions with a mean and standard deviation of $1/\Lambda$, where T_c is the arrival time of the first ray of the cth cluster. In the same way, within each cluster, the consecutive ray times of arrival $\tau_{c,r+1} - \tau_{c,r}$ can be modeled by exponentially distributed random variables with a mean and standard deviation of $1/\lambda$, where $\tau_{c,r}$ is the delay of the rth ray within the cth cluster relative to the first path arrival time of that cluster.

Contrary to the model proposed in [35], the multipath gain magnitude distribution of the IEEE 802.15.3a channel model is log–normal. Since this channel modeled a multipath propagation in an indoor environment, the channel response can be considered invariant for the duration of several symbols, that is, the slow fading condition applies. Moreover, the fading is assumed to be independent for each cluster and each ray within any cluster. Therefore, the channel impulse response per cluster, per ray, $h_{c,r}(t)$, can be expressed as

$$h_{c,r}(t) = \alpha_{c,r}\delta(t - T_c - \tau_{c,r}) \tag{4.44}$$

where the normalized multipath gain

$$\alpha_{c,r} = \frac{\sqrt{\Omega_0}\xi_c\beta_{c,r}e^\phi e^{j\theta_{c,r}}}{\sqrt{\sum_{g=0}^{\infty}\sum_{f=0}^{\infty}(\xi_g\beta_{g,f})^2}} \tag{4.45}$$

includes amplitudes ξ_c and $\beta_{c,r}$, and phase $\theta_{c,r}$ components corresponding to the cth cluster and the rth ray, a normalization term $\sqrt{\Omega_0/\sum_{g=0}^{\infty}\sum_{f=0}^{\infty}(\xi_g\beta_{g,f})^2}$ such that

$$\sum_{c=0}^{\infty}\sum_{r=0}^{\infty}\alpha_{c,r}^2 = \Omega_0 e^{2\phi} \tag{4.46}$$

and shadowing effects characterized by a $(0, \sigma_s)$ normally distributed random variable ϕ. The statistically independent uniform random variables $\theta_{c,r} = \{0, \pi\}$ accounts for the reflections of the propagated signal. The amplitude component ξ_c accounts for the fading gain of the cth cluster, and $\beta_{c,r}$ is defined as the fading gain of the rth ray within the cth cluster. Any statistically independent random variable defined as $\xi_c\beta_{c,r}$ is the product of two independent random variables such that ξ_c is (μ_c, σ_1) log–normally distributed and $\beta_{c,r}$ is $(m_{c,r}, \sigma_2)$ log–normally distributed, with

$$\mu_c = \frac{\ln(\Omega_0)}{2} - \sigma^2 - \frac{T_c}{2\Gamma}$$

$$m_{c,r} = j\theta_{c,r} - \frac{\tau_{c,r}}{2\gamma}$$

$$\sigma = \sqrt{\sigma_1^2 + \sigma_2^2}$$

Ω_0 represents the mean energy of the first path of the first cluster. This average energy is assumed to be equal to one, i.e. $\Omega_0 = 1$. Thus, the multipath gain magnitude $\alpha_{c,r}$ can be expressed as

$$\alpha_{c,r} = \frac{\exp[-(T_c/2\Gamma) + Y_c - (\tau_{c,r}/2\gamma) + X_{c,r} - \sigma^2 + \phi + j\theta_{c,r}]}{\sqrt{\sum_{g=0}^{\infty} \sum_{f=0}^{\infty} (\xi_g \beta_{g,f})^2}} \tag{4.47}$$

where Y_c are $(0, \sigma_1)$ and $X_{c,r}$ are $(0, \sigma_2)$ normally distributed random variables, respectively. According to the nature of T_c and $\tau_{c,r}$, one can define a $1/(\Lambda\Gamma)$ exponentially distributed random variable R_c as

$$\frac{T_c}{\Gamma} = R_c + \frac{T_{c-1}}{\Gamma} \tag{4.48}$$

where T_{-1}/Γ is equal to zero. In the same way, a $1/(\lambda\gamma)$ exponentially distributed random variable $S_{c,r}$ is defined as

$$\frac{\tau_{c,r+1}}{\gamma} = S_{c,r+1} + \frac{\tau_{c,r}}{\gamma} \tag{4.49}$$

where $\tau_{c,0}/\gamma$ and $S_{c,0}$ are equal to zero. Next, with Equations (4.48) and (4.49), we can rewrite the multipath gain magnitude introduced in Equation (4.47) as

$$\alpha_{c,r} = \frac{e^{j\theta_{c,r} + \phi} e^{Y_c + X_{c,r}} \prod_{i=0}^{c} \prod_{j=0}^{r} e^{-(R_i/2 + S_{c,j}/2)}}{\sqrt{\sum_{g=0}^{\infty} \sum_{f=0}^{\infty} e^{2(Y_g + X_{g,f})} \prod_{i=0}^{g} \prod_{j=0}^{f} e^{-(R_i + S_{g,j})}}} \tag{4.50}$$

Since cluster and ray amplitude decays are proportional to $e^{T_c/2\Gamma}$ and $e^{-\tau_{c,r}/2\gamma}$, respectively, for large values of T_c and $\tau_{c,r}$, or equivalently for large values of indices c and r, one may expect that $\alpha_{c,r} \to 0$. Then, we define the average number of rays per cluster n_c, such that $\tau_{c,n_c-1} < 10\gamma$ and $\tau_{c,n_c} \geq 10\gamma$. In the same way, we define the average number of clusters N, such that $T_{N-1} < 10\Gamma$ and $T_N \geq 10\Gamma$ [50]. Finally, the mathematical definition of the channel impulse response in terms of n_c, N, Equations (4.44), and (4.50) can be given by

$$h(t) = \frac{\sum_{c=0}^{N-1} \sum_{r=0}^{n_c-1} e^{j\theta_{c,r}} e^{Y_c + X_{c,r}} \prod_{i=0}^{c} \prod_{j=0}^{r} e^{-(R_i/2 + S_{c,j}/2)}}{\sqrt{\sum_{g=0}^{N-1} \sum_{f=0}^{n_g-1} e^{2(Y_g + X_{g,f})} \prod_{i=0}^{g} \prod_{j=0}^{f} e^{-(R_i + S_{g,j})}}} e^{\phi} \delta(t - T_c - \tau_{c,r}) \tag{4.51}$$

For simplicity and convenience, the channel impulse response of the IEEE 802.15.3a channel model can also be expressed with Equation (4.44) as

$$h(t) = \sum_{d=0}^{N_m-1} \alpha_d \delta(t - \tau_d) \tag{4.52}$$

where

$$N_m = \sum_{c=0}^{N-1} n_c \tag{4.53}$$

is the average total number of rays,

$$d = \sum_{g=0}^{c} n_{g-1} + r, n_{-1} = 0 \qquad (4.54)$$

$d \in D$, $D = \{0, 1, \ldots, N_m - 1\}$ is the set of multipath component indexes, $\alpha_d = \alpha_{c,r}$, and

$$\tau_d = T_c - \tau_{c,r} \qquad (4.55)$$

The IEEE 802.15.3a channel model is composed of four different channel scenarios defined according to seven key parameters:

- Λ: cluster arrival rate;

- λ: ray arrival rate, i.e. the arrival rate of path within each cluster;

- Γ: cluster decay factor;

- γ: ray decay factor;

- σ_1: standard deviation of cluster log–normal fading term (dB);

- σ_2: standard deviation of ray log–normal fading term (dB);

- σ_s: standard deviation of log–normal shadowing term for total multipath realization (dB).

These parameters are obtained as the best effort to match the most important characteristics of the channel, such as:

- the mean and rms excess delay;

- the number of multipath components;

- the power decay profile.

Four different channel characteristics have been obtained based on many experiments and much measurement data. These channel characteristics depend on the average distance between the transmitter and the receiver, and the type of transmission, i.e. either LOS or non-LOS (NLOS) transmissions. The first model, called the channel model type 1 (CM1) scenario, corresponds to very short distances, i.e. 0 to 4 m, and LOS transmissions. The channel model type 2 (CM2) scenario is defined for the same range, but with a NLOS antenna configuration. CM3 and CM4 are defined for a NLOS antenna configuration and greater transmission distances, i.e. 4 to 10 m for CM3 and over 10 m for CM4.

The main parameters of the four channel scenarios are listed in Table 4.1, where $\mathrm{NP}_{10\ \mathrm{dB}}$ represents the number of paths within 10 dB of the peak, NP (85%) stands for the number of paths capturing 85% of the energy. These parameters are obtained

Table 4.1 The main parameters of the four channel scenarios composing the IEEE 802.15.3a channel model.

	CM1	CM2	CM3	CM4
Target channel characteristics				
Mean excess delay [ns]	5.0	10.38	14.18	
rms delay [ns]	5.28	8.03	14.28	25
$NP_{10\,dB}$			35	
NP (85%)	24	36	84	
Model parameters				
Λ [1/ns]	0.0233	0.4	0.0667	0.0667
λ [1/ns]	2.5	0.5	2.1	2.1
Γ	7.1	5.5	14	24
γ	4.3	6.7	7.9	12
σ_1	0.3907	0.3907	0.3907	0.3907
σ_2	0.3907	0.3907	0.3907	0.3907
σ_s	0.3454	0.3454	0.3454	0.3454
Model characteristics				
Mean excess delay [ns]	5.1	9.6	15.3	28.5
RMS delay [ns]	5	8	15	25
$NP_{10\,dB}$	17.7	17.7	34.3	53.4
NP (85%)	23.4	34.3	80.6	191.3
Channel energy mean [dB]	0.0	−0.3	−0.1	0.2
Channel energy standard [dB]	3	3.1	3.1	3.0

by using the Matlab program, provided in [50], generating 1000 independent channel realizations of the time-continuous channel impulse response, and collecting the average result for each parameter.

Figure 4.13 illustrates the differences of the characteristics of each channel scenario. The average delay profile for each of them, averaged over 100 independent realizations, is shown. Channel models CM1 and CM2 present similarity in the way their multipath components are spread in time. Otherwise, the difference between LOS and NLOS transmissions is clearly visible between CM1 and CM2, since in CM2 the group of the strongest multipath components is delayed by about 5 ns in comparison with the same group in CM1. Finally, increasing the distance between the transmit and receive antennas implies a spreading in time of the multipath components, as seen for CM3 and CM4.

Up to the present time, most of the channel models proposed for UWB communications support the clustered ray arrival assumption with different multipath amplitude distributions. As far as the UWB path loss is concerned, general propagation physics approaches are valid, i.e. a longer distance between transmitter and receiver elements implies a lower amount of energy received. This amount of transmitted energy received

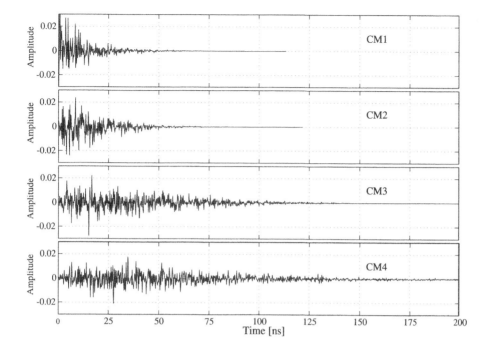

Fig. 4.13 Average delay profiles for each channel scenario, CM1–CM4.

at any distance d of the transmitter can be computed by a propagation loss calculation. Using the large amount of measurement data collected in the process of channel modeling, the UWB free space path loss model has been characterized such that $\text{PL}(d) = (d/d_0)^{-n}$, where d_0 is a reference distance, i.e. usually $d_0=1$ m, and $n = 2$. This model exhibits a breakpoint, and can be characterized by a dual-slope model.

The mathematical definition of the IEEE 802.15.3a channel impulse response given in Equation (4.51) is the normalized-to-unit-energy definition of the channel impulse response given in the core of the report [50]. Actually, the definition in Equation (4.51) corresponds exactly to the definition used in the Matlab program to generate any channel impulse response and to simulate this channel model. This proper definition leaves the multipath gain magnitude $\alpha_{c,r}$, i.e. Equation (4.50), expressed in terms of a sum of a mixture of log–normal random variables and an exponential of exponential random variables. It should be possible to prove that any multipath gain magnitude $\alpha_{c,r}$ can be synthesized into a single random variable which follows a well-approximated log–normal distribution. Hence, we can develop an analytical method to obtain the mean and the standard deviation of this random variable according to the channel parameters given for each channel scenario in Table 4.1, as shown in [52].

4.6 SUMMARY

In this chapter the propagation aspects of UWB signals in indoor and indoor–outdoor environments were discussed. We started with a simplified multipath channel model that was not unique to UWB systems, but could be an appropriate approach. The primary parameters that are important to characterize the indoor channel, including the number of resolvable multipath components, multipath delay spread, multipath intensity profile, multipath amplitude-fading distribution, and multipath arrival times were briefly described.

As a second approach to UWB channel characterization we considered the path loss model. Various effects and phenomena involved, such as free space loss, refraction, reflection, diffraction, clutter, aperture–medium coupling loss, and absorption, were defined and explained. An example of a free space path loss consideration was also investigated and a simple two-ray UWB propagation mechanism was thoroughly studied.

As the third approach, the statistical frequency domain autoregressive model was explained. The target of a frequency domain model is to develop a statistical representation of the UWB channel with a minimum number of parameters to regenerate the measured channel behavior accurately in computer simulations. The higher the complexity of the model, the closer the statistical resemblance to that of measured data.

Finally, the IEEE 802.15.3a channel model with the main features of the Saleh–Valenzuela channel model, where multipath rays arrive in clusters and their amplitudes follow a double-exponential decay, was explained. Each multipath gain magnitude coefficient of this frequency-selective channel model is assumed to be a log–normally distributed random variable. This model is designed for baseband signalling. An analytical description of the IEEE 802.15.3a channel model was also presented.

Problems

Problem 1. What are the important parameters involved in multipath channel characterization of UWB systems? Briefly describe each one.

Problem 2. Discuss the importance of Doppler spread in UWB channel modeling.

Problem 3. Why can use of the Poisson process be a proper way for statistically modeling the arrival times of different paths?

Problem 4. Investigate the main differences between the path loss models proposed by Cramer et al. [38] and Ghassemzadeh et al. [39].

Problem 5. Derive Equation (4.25).

Problem 6. Describe the distance and frequency dependence behavior of the two-ray path loss model using Equation (4.29).

Problem 7. Explain how the transmitted and received signals of Figures 4.11(b) and 4.11(c) are related.

Problem 8. What is an AR model?

Problem 9. It has been shown that the largest pole of an autoregressive UWB model is very close to the unit circle. How does this property affect the performance of the system?

5

UWB communications

In this chapter we will look at the use of UWB wireless communications. From the treatment of individual pulse shaping and generation, which was introduced in Chapter 2, we now move on to examine various communications concepts. Particular attention will be paid to modulation methods including pulse position modulation, bi-phase modulation, orthogonal pulse modulation, and their combinations. Sequences of individual pulses onto pulse streams will be presented. Receiver design and pulse detection will be examined.

We also move from a single-user environment to examine multiple access techniques for UWB communications. The capacity of the wireless UWB channel will also be examined. The effect of UWB on existing wireless communication methods, such as the IEEE 802.11 WLAN standards and Bluetooth, is shown. Several methods to prevent interference from these narrowband systems to UWB will also be examined.

Finally, an important discussion on the relative merits and demerits of UWB as a communication method with other wideband communication techniques, such as CDMA and orthogonal frequency division multiplexing (OFDM), is undertaken.

5.1 INTRODUCTION

Communication can generally be defined as the transmission of information from a source to a recipient. In this chapter we make our definition of communication much narrower, by restricting ourselves to wireless communication of digital data streams of information using extremely short pulses. We deliberately ignore the kind of information contained in the digital data stream and do not use *media access control* (MAC) protocols, coding schemes, or retransmission schemes to reduce errors.

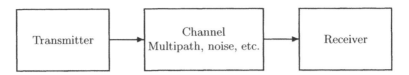

Fig. 5.1   Model of a general communications system.

We concentrate on what is commonly known as the *physical layer* in the International Standards Organization (ISO) protocol stack.

A general model of a communication system is shown in Figure 5.1. The three basic elements are as follows:

- the *transmitter*, whose primary task is to group the digital data stream into symbols, to map these symbols onto an analog waveform, and then to transmit them to the air through an antenna;

- the *channel*, which represents the effect of traveling through space, including reflections and distortions as the electromagnetic pulses impinge on other objects;

- the *receiver*, which collects the electromagnetic energy from the antenna, takes the extremely weak signal, reconstructs the pulse shape, and maps it to the appropriate symbols and then to the binary bitstream.

In this chapter we examine receiver and transmitter structures in more detail, focusing on the basic communication aspects, such as modulation. Channel effects have already been covered in Chapter 4.

5.2   UWB MODULATION METHODS

As we saw in Chapter 1, one single UWB pulse does not contain information by itself. We must add digital information to the analog pulse, by means of *modulation*. In UWB systems there are several basic methods of modulation, and we examine each in detail.

As a helpful categorization of modulation methods, we define two basic types for UWB communication. These are shown in Figure 5.2 as *time-based* techniques and *shape-based* techniques.

By far the most common method of modulation in the literature is *pulse position modulation* (PPM) where each pulse is delayed or sent in advance of a regular time scale. Thus, a binary communication system can be established with a forward or backward shift in time. By specifying specific time delays for each pulse, an *M*-ary system can be created.

Another common method of modulation is to invert the pulse, that is, to create a pulse with opposite phase. This is known as *bi-phase modulation* (BPM).

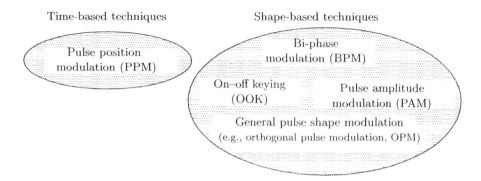

Fig. 5.2 Division of different modulation methods for UWB communications.

An interesting modulation technique is *orthogonal pulse modulation* (OPM), which requires special pulse shapes to be generated which are orthogonal to each other.

Other well-known techniques for modulation are available. For example, *on–off keying* (OOK) is a modulation technique where the absence or presence of a pulse signifies the digital information of '0' or '1', respectively. *Pulse amplitude modulation* (PAM) is a technique where the amplitude of the pulse varies to contain digital information.

Furthermore, some traditional modulation techniques are not available to us. For example, the widely used *frequency modulation* (FM) is difficult to apply to UWB, since each pulse contains many frequency elements making it difficult to modulate. Note that this should not be confused with *frequency division multiplexing* (FDM) which is an entirely different technique to separate communication channels based on larger blocks of frequency (discussed later).

Let us examine each of these possible modulation techniques in turn. First, we examine the two most common techniques, PPM and BPM. A simple comparison of these two modulation methods is shown in Figure 5.3. In Figure 5.3(a) an unmodulated pulse train is shown for comparison. As an example of PPM, the pulse representing the information '1' is delayed in time (i.e. the pulse appears to be moved in position to the right). The pulse representing the information '0' is sent before the nonmodulated pulse (i.e. the pulse appears to be moved in position to the left) in Figure 5.3(b). In BPM the inverted pulse represents a '0' while the uninverted pulse represents a '1'. This is clearly illustrated in Figure 5.3(c).

5.2.1   PPM

As mentioned previously, the important parameter in PPM is the delay of the pulse. That is, by defining a basis pulse with arbitrary shape $p(t)$, we can modulate the data by the delay parameter τ_i to create pulses s_i, where t represents time,

$$s_i = p(t - \tau_i) \tag{5.1}$$

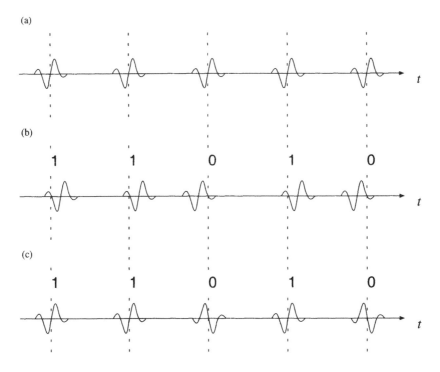

Fig. 5.3 Comparison of (a) an unmodulated pulse train, (b) PPM, and (c) BPM methods for UWB communication.

As an example, we can let $\tau_1 = -0.75$, $\tau_2 = -0.25$, $\tau_3 = 0.25$, and $\tau_4 = 0.75$, to create a 4-ary PPM system. The four pulse shapes become

$$
\begin{aligned}
s_1 &= p(t + 0.75) \\
s_2 &= p(t + 0.25) \\
s_3 &= p(t - 0.25) \\
s_4 &= p(t - 0.75)
\end{aligned}
\tag{5.2}
$$

The advantages of PPM mainly arise from its simplicity and the ease with which the delay may be controlled. On the other hand, for the UWB system extremely fine time control is necessary to modulate pulses to sub-nanosecond accuracy.

5.2.2 BPM

BPM can be defined as a kind of shape modulation. Since phase in a sinusoidal communication system is associated with the delay of a sine wave, the overuse of the term phase in UWB can be confusing. However, the use of BPM has become common in the UWB literature, so we continue to use it here. BPM is easily understood as the inversion of a particular pulse shape; therefore, we take the equation

$$s_i = \sigma_i p(t), \qquad \sigma_i = 1, -1 \qquad (5.3)$$

to create a binary system based on inversion of the basis pulse $p(t)$. The parameter σ is often known as the *pulse weight*, but here we will refer to it as the shape parameter. For a binary system the two resultant pulse shapes s_1, s_2 are defined simply as $s_1 = p(t)$ and $s_2 = -p(t)$.

One of the reasons for the use of BPM, especially in comparison with PPM (which is a mono-phase technique), is the 3 dB gain in power efficiency. This is simply a function of the type of modulation method. That is, BPM is an *antipodal modulation method*,[1] whereas PPM, when separated by one pulse width delay for each pulse position, is an *orthogonal modulation method*. The interested reader should refer to any digital communications textbook for details (see [53]).

A simple example can illustrate this advantage of BPM. Since PPM must always delay pulses, in the limit when pulses are transmitted continuously PPM must always 'waste' the time when pulses are not transmitted. If PPM delays by one pulse width, then BPM can send twice the number of pulses and, thus, twice the information, so as to achieve a system which, given all other things being equal, has twice the data rate.

Another benefit of using BPM is that the mean of σ is zero. This has the important benefit of *removing* the comb lines or spectral peaks that were discussed in Chapter 1, without the need for 'dithering'. This of course assumes that transmitted bits are equally likely; however, this is a common and reasonable assumption in most digital communication systems. According to McCorkle [54], BPM in UWB presents several other benefits:

> First, it exhibits a peak-to-average power ratio of less than 8 dB. Thus, an implementation using bi-phase does not require any external snap-recovery or tunnel diodes or power-amplifier circuitry. Instead, it can be driven directly from a low-voltage high-speed complementary metal-oxide-semiconductor (CMOS) IC.
>
> Finally, for reasons of clocking, bi-phase modulation has reduced jitter requirements. In PPM, the clocking path must include elements to accurately control arbitrary time positions on a fast (pulse-to-pulse) basis. This control requires a series of wide-bandwidth circuits where jitter accumulates. But a bi-phase system needs only a stable, low-phase-noise clock as the pulses occur on a constant spacing. Synchronization circuits can be narrowband so that they do not add significant jitter. As a result, less power and real estate are needed to implement the required circuits.

5.3 OTHER MODULATION METHODS

Although the previously discussed PPM and BPM constitute the major approaches to modulation in UWB communication systems, other approaches have been proposed.

---

[1] Antipodal means opposite.

A simple diagram outlining these alternative modulation approaches is shown in Figure 5.4.

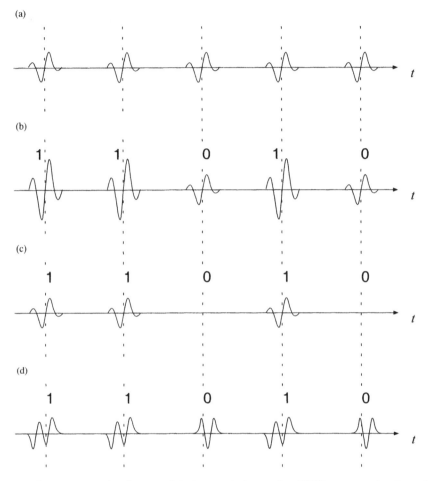

Fig. 5.4 Comparison of other modulation techniques for UWB communication: (a) an unmodulated pulse train; (b) PAM; (c) OOK, and (d) OPM.

In Figure 5.4(a) an unmodulated pulse train is shown for comparison. In Figure 5.4(b) an example of PAM is given where a pulse with large amplitude represents a '1' and smaller amplitude represents a '0'. In Figure 5.4(c) OOK modulates data with the presence of a pulse representing '1' and an absence representing '0'. Figure 5.4(d) shows an example of an OPM where a binary '1' is represented by a modified Hermitian pulse of order 3 and a binary '0' is represented by a modified Hermitian pulse of order 2 [17].

5.3.1 OPM

Of the three unconventional modulation techniques, OPM is simply a subset of general pulse shape modulation (PSM) with the property that the pulse shapes are orthogonal to each other. The advantage of using orthogonal pulses is not strictly related to the modulation, but rather to the multiple access method, which is discussed later in this chapter. However, since arbitrary PSM is not interesting in either a theoretical or practical sense,[2] here we consider OPM as an interesting subset of the general case of PSM.

In narrowband sinusoidal communication, orthogonal sine and cosine functions form the basis for communication. In UWB we can design different pulse shapes that have the property of being orthogonal to each other. Unfortunately, a simple pulse shape parameter σ is inadequate to describe the set of pulses which we may encounter, and here we simply label each pulse as a general $p_1, p_2, \ldots, p_i$ and assume that pulses are designed so as to be orthogonal (see [17] and Chapter 6 for examples on how to design orthogonal pulse shapes):

$$s_i = \{p_1, p_2, \ldots, p_i\} \tag{5.4}$$

The three modulation techniques presented previously, PPM, BPM, and OPM, have been proposed for use in UWB communications. As far as the authors are aware, no serious attempt has been made to use either PAM or OOK for UWB; however, for completeness here their possible use is discussed below.

5.3.2 PAM

PAM for UWB can be represented as

$$s_i = \sigma_i p(t), \qquad \sigma_i > 0 \tag{5.5}$$

where the pulse shape parameter σ takes on positive values greater than zero. As an example, we can set $\sigma_i = 1, 2$ and obtain the binary pulse set $s_1 = p(t)$, $s_2 = 2p(t)$.

In general, amplitude modulation is not the preferred way for most short-range communication. The major reasons for this include the fact that, in general, an amplitude-modulated signal which has a smaller amplitude is more susceptible to noise interference than its larger-amplitude counterpart. Furthermore, more power is required to transmit the higher-amplitude pulse.

In sinusoidal systems, amplitude-modulated systems are usually characterized by a relatively low bandwidth requirement and power inefficiency in comparison with angle modulation schemes. Thus, the major advantage (low bandwidth) can be seen to be anti-ethical to UWB, and in most UWB applications power efficiency is of high importance.

---

[2]This is simply because additional circuitry, memory, and so on is needed to generate the arbitrary pulse shapes. If there is no reason for adding complexity to the system, we should not do it.

5.3.3 OOK

OOK for UWB can be characterized as a type of PSM where the shape parameter s is either 0 or 1,

$$s_i = \sigma_i p(t), \qquad \sigma_i = 0, 1 \tag{5.6}$$

For example, the 'on' pulse is created when $\sigma_i = 1$ and the 'off' pulse when $\sigma_i = 0$; thus, $s_1 = p(t)$ and $s_2 = 0$.

The major difficulty of OOK is the presence of multipath, in which echoes of the original or other pulses make it difficult to determine the absence of a pulse. OOK is also a binary modulation method, similar to BPM, but it cannot be extended to an M-ary modulation method, as can PPM, PAM, and OPM.

5.3.4 Summary of UWB modulation methods

In this sub-section we conclude the discussion of modulation methods for UWB communications with Table 5.1, which summarizes the advantages and disadvantages of each of the modulation methods.

Table 5.1 Advantages and disadvantages of various modulation methods.

Modulation methods	Advantages	Disadvantages
PPM	Simplicity	Needs fine time resolution
BPM	Simplicity, efficiency	Binary only
OPM	Orthogonal for multiple access	Complexity
PAM	Simplicity	Low noise immunity
OOK	Simplicity	Binary only, low noise immunity

5.4 PULSE TRAINS

We examined the creation of single pulses in Chapter 2. In this chapter we have looked at sets of pulses which are used for the modulation of digital information onto analog pulse shapes. We now turn our attention to sequences of pulses, called pulse trains, which will be able to transmit much larger volumes of information than a single set of pulses. In general, an unmodulated pulse train $s(t)$ with a regular pulse output can be written as

$$s(t) = \sum_{n=-\infty}^{\infty} p(t - nT) \tag{5.7}$$

where T is the period or the pulse-spacing interval and $p(t)$ is the basis pulse.

The effects of changing the pulse duration and repetition rate of each pulse have been examined and the results are as follows:

- Increasing the pulse rate in the time domain increases the magnitude in the frequency domain (i.e. the pulse rate influences the magnitude of the spectrum).

- The lower the pulse duration in the time domain, the wider the spectral width (i.e. the pulse duration determines spectral width).

- A random pulse-to-pulse interval produces a much lower peak magnitude spectrum than a regular pulse-to-pulse interval since the frequency components are unevenly spread over the spectrum and the addition of magnitude is less effective. Therefore, the pulse-to-pulse interval controls the separation of spectral components.

5.4.1 Gaussian pulse train

As an example of a pulse train, let us consider the Gaussian doublets of Figure 2.2 with PPM.

As briefly mentioned in Chapter 1, assuming PPM is the modulation method, there is the problem of spectral peaks when a regular pulse train is used. These energy spikes can cause interference with other RF systems at short range and limit the amount of useful energy transmitted. One method to overcome these spectral peaks is to 'dither' the transmitted signal by adding a random offset to each pulse, removing the common spectral components. However, when we attempt communication with this model the random offset is unknown at the receiver, making it extremely difficult to acquire and track the transmitted UWB signal. Another method with similar random properties, but using a known sequence, is to use *pseudo-random noise* (PN) codes to add an offset to the PPM signal. Since these codes are known and easily reproducible at the receiver, the problem for the receiver becomes mostly acquisition of the signal, but tracking is made much easier.

5.4.2 PN channel coding

The use of a PN time shift has other benefits besides just reducing the spectral peaks resulting from regular pulse emissions. Since the PN code is a channel code it can be used as a multiple access method to separate users in a similar manner to the CDMA scheme. By shifting each pulse at a pseudo-random time interval the pulses appear to be white background noise to users with a different PN code. Furthermore, the use of PN codes makes data transmission more secure in a hostile environment.

The impact of PN time offsets on energy distribution in the frequency domain is illustrated in Figure 5.5. Here, the basis pulse $p(t)$ used for transmission of the signals is assumed to be the Gaussian doublet of Figure 2.2, that is, $y_{g3}(t)$. A sequence of pulses is formed,

$$s(t) = \sum_{n=0}^{N_t-1} y_{g3}(t - t_{d_n}) \tag{5.8}$$

where $y_{g3}(t)$ has been defined in Equation (2.7), N_t is the number of doublets, and t_{d_n} is the time delay associated with the pulse number n. This time delay is related

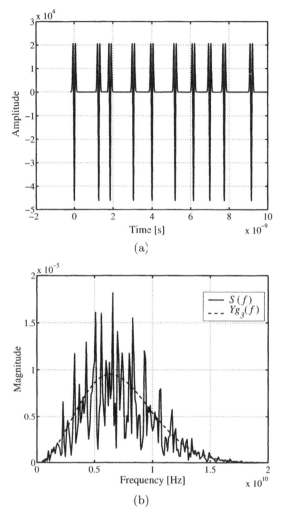

Fig. 5.5 A train of Gaussian doublets in the (a) time and (b) frequency domains.

to the input code and can be calculated in different ways, for example,

$$t_{d_n} = nT_f + T_{p_n} + T_{c_n} \tag{5.9}$$

where T_f is the pulse repetition time, T_{c_n} is the random time shift related to the PN code sequence, and T_{p_n} is the time shift from the PPM scheme. In practice, T_f is much greater than T_{p_n} or T_{c_n}.

Using the Fourier transform of $y_{g_3}(t)$ indicated in Equation (2.14) we can conclude that the spectrum of $s(t)$ is

$$S(f) = \sum_{i=0}^{N_t-1} K_3 \tau \sqrt{\pi} (j2\pi f)^2 e^{-(\pi\tau f)^2} e^{-j2\pi f t_{d_n}} \qquad (5.10)$$

where K_3 is calculated based on Equation (2.11) for $E_3 = 0.1$.

Example 5.1

Using Equation (5.8), plot the time domain function of a Gaussian doublet train. Use $N_t = 10$ and let the energy of each doublet be equal to 0.1, which gives a total energy of unity.

Using Equation (5.10), plot the frequency domain function of the Gaussian doublet train. Compare the result with the frequency domain representation of a single doublet $y_{g_3}(t)$.

Solution

The curves are plotted in Figures 5.5(a) and 5.5(b).

5.4.3 Time-hopping PPM UWB system

We can now combine the techniques introduced to build a simple UWB transmitter. We will use a time-hopping code and binary PPM, with a single reference pulse shape $p(t)$. This system is perhaps the most common in the literature (e.g., see [10]). It only requires a single template pulse for reception, and most of the complexity of this system resides in providing accurate timing for the generation of the transmitted sequence and subsequent reception.

In Figure 5.6 we show the output of this simple UWB transmitter. We describe it here for the single-user case, but we can extend it easily and simply for the multi-user case by using different time-hopping codes, which in general will be PN codes.

First, we note that there is one pulse transmitted in each frame of time T_f. The *pulse repetition frequency* (PRF), as described previously, is

$$\text{PRF} = \frac{1}{T_f} \qquad (5.11)$$

The frame time should be at least long enough to overcome the delay spread of the channel, which, as described in Chapter 4, is generally of the order of hundreds of nanoseconds for an indoor environment, in order to avoid interference from reflected pulses. Thus, the frame time will be of the order of 1000 times the actual pulse width. The unmodulated pulse stream is represented as

$$s(t) = \sum_{n=-\infty}^{\infty} p(t - nT_f) \qquad (5.12)$$

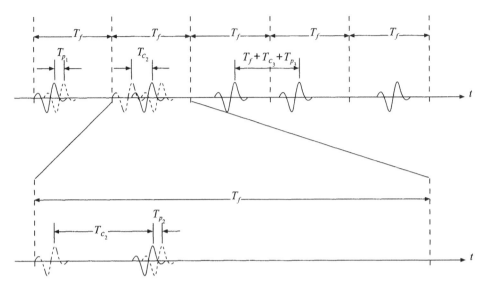

Fig. 5.6 A time-hopping, binary PPM system output.

To modulate the data we add a small shift in the pulse position T_{p_n}, either forward T_{p_0} or backward T_{p_1}, to represent the binary data stream. Often $T_{p_0} = -T_{p_1}$. The modulated data stream becomes

$$s(t) = \sum_{n=-\infty}^{\infty} p\left(t - nT_f - T_{p_n}\right) \qquad (5.13)$$

To avoid spectral lines and to provide a means of distinguishing users we finally add a time shift based on a time-hopping code T_{c_n}, where the code repeats after a certain interval. The final output of the time-hopping, PPM signal is then given by

$$s(t) = \sum_{n=-\infty}^{\infty} p\left(t - nT_f - T_{p_n} - T_{c_n}\right) \qquad (5.14)$$

5.5 UWB TRANSMITTER

A general UWB transmitter block diagram is shown in Figure 5.7. First, meaningful data are generated by applications that are quite separate from the physical layer transmitter. Applications might be an e-mail client or a web browser on a personal computer, a calendar application on a personal digital assistant (PDA), or the digital stream of data from a DVD player. From the perspective of the physical layer the data may be anything at all. This part of the wireless device is often called the 'back end'. This terminology is not immediately apparent, but it is common to refer to it as from the receiver's point of view.

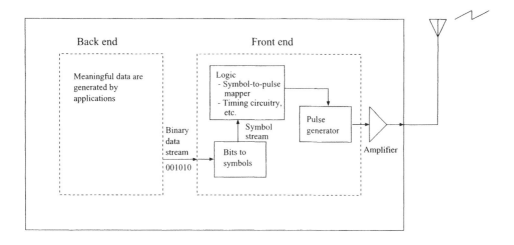

Fig. 5.7   A general UWB transmitter block diagram.

This binary information stream is then passed to the 'front end', which is the part of the transmitter which we are concerned about. If higher modulation schemes are to be used the binary information should be mapped from bits to symbols, with each symbol representing multiple bits. These symbols are then mapped to an analog pulse shape. Pulse shapes are generated by the pulse generator. Precise timing circuitry is required to send the pulses out at intervals which are meaningful. If PPM is employed the timing must be even more precise, usually less than one pulse width.

Pulses can then be optionally amplified before being passed to the transmitter. In general though, to meet power spectral requirements, a large gain is typically not needed and may be omitted.

Although this is an extremely simplistic transmitter model, which omits any forward error-correcting scheme, it serves the purpose to show that UWB transmitters can be quite simple. This is to be compared with other wireless transmitters, such as OFDM. See Figure 5.17 for comparison.

5.6   UWB RECEIVER

A general UWB receiver block diagram is shown in Figure 5.8. The receiver performs the opposite operation of the transmitter to recover the data and pass the data to whatever 'back end' application may require it.

There are two major differences between the transmitter and the receiver. The first is that the receiver will almost certainly have an amplifier to boost the signal power of the extremely weak signals received. The second is that the receiver must

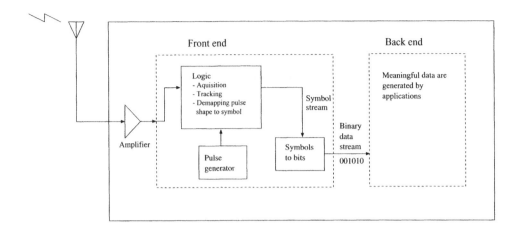

Fig. 5.8 A general UWB receiver block diagram.

perform the functions of *detection* or *acquisition* to locate the required pulses amongst the other signals and then to continue *tracking* these pulses to compensate for any mismatch between the clocks of the transmitter and the receiver.

Communication requires both the transmission and reception of signals. We have mostly concentrated on the wireless transmission side up to this moment. We will now focus on the detection of pulses, that is, the acquisition and tracking of the pulse trains.

5.6.1 Detection

Having generated a signal with the desired spectral features, it is also necessary to have an optimal receiving system. The optimal receive technique, the technique often used in UWB, is a correlation receiver, usually known as a *correlator*. A correlator multiplies the received RF signal by a *template waveform* and then integrates the output of that process to yield a single DC voltage. This multiply-and-integrate process occurs over the duration of the pulse and is performed in less than a nanosecond. With the proper template waveform the output of the correlator is a measure of the relative time positions of the received monocycle and the template.

If we assume PPM as the modulation method the correlator is an optimal early/late detector. As a very simple example, when the received pulse is one-quarter of a pulse early the output of the correlator is +1, when it is one-quarter of a pulse late the output is −1, and when the received pulse arrives centered in the correlation window the output is zero.

It is critical to note that the mean value of the correlator is zero. Thus, for in-band noise signals received by a UWB radio the correlator's output has an average value of zero. Moreover, the standard deviation or rms of the correlator output is related to the power of those in-band noise signals.

5.6.2 Pulse integration

When a monocycle is buried in the noise of other signals, it is extremely difficult to detect a single UWB pulse and the confidence with which we can receive the transmitted information is low. However, by adding together multiple correlator samples (i.e. multiple pulses), it becomes possible to receive transmitted signals with much higher confidence. This process is called *pulse integration*. Through pulse integration, UWB receivers can acquire, track, and demodulate UWB transmissions that are significantly below the noise floor. The measure of a UWB receiver's performance in the face of in-band noise signals is called the *processing gain*.

5.6.3 Tracking

Tracking is the process by which the receiver must continually check to see whether the pulses are arriving at the expected time and, if not, to adjust that time. A simple example will serve to show the process. Assume that the transmitter and receiver start with their clocks synchronized. As time passes the effects of heat and differences in manufacture cause one of the clocks or oscillators to become slightly faster. If this difference is not corrected, eventually the receiver will not be able to correctly demodulate the pulses. The time drift at sub-nanosecond orders must be vigilantly watched, in particular.

5.6.4 Rake receivers

As discussed in Chapter 4, the wireless channel suffers from *multipath*, where reflections and other effects of the channel cause multiple copies of the transmitted pulse to appear at the receiver. If a *rake receiver* is used, these extra pulses can be used to improve reception at the cost of increased receiver complexity. The increased complexity comes from the additional circuitry required to track multiple pulses and demodulate them. The name *rake* receiver comes from the fact that the delay profile of the received pulse looks like an upturned garden rake.

5.7 MULTIPLE ACCESS TECHNIQUES IN UWB

Up to now we have implicitly assumed that there has only been one user using the UWB system at any one time. Although consumer UWB systems are in their infancy, we must consider how best to design a UWB system where there is more than one user.

Naturally, all traditional multiple access methods should be considered; however, here we consider time, frequency, code, and space division multiple access. As an interesting special case for UWB, we look at orthogonal pulses as a novel multiple access technique.

5.7.1 Frequency division multiple access UWB

A common multiple access technique in narrowband communication is to divide users up based on frequency bandwidth. This is known as frequency division multiple access (FDMA). Each user uses a different carrier frequency to transmit and receive on.

In UWB, FDMA is achieved by using pulses which have a narrower bandwidth than the total available bandwidth; however, they are still extremely broadband. Channelization can be achieved by multiplying by a sinusoidal carrier. The total effect can be thought of as an extremely broadband OFDM system.

5.7.2 Time division multiple access

In time division multiple access (TDMA), each user uses the same codes and the same bandwidth; however, a different time offset is needed to avoid interference. In general, this requires that all users be synchronized, which is not an easy task as the number of users increases. In general, this technique would only be applied to the downlink (from a central base station) to mobile users.

5.7.3 Code division multiple access

One multiple access technique possible in UWB is to assign a different spreading code to each user. This is known as code division multiple access (CDMA).

As an example, let us take Equation (5.14) and modify it so that we can separate k users out by different code. We refer to the kth user's output from the transmitter as

$$s^{(k)}(t) = \sum_{n=-\infty}^{\infty} p\left(t - nT_f - T_{p_n}^{(k)} - T_{c_n}^{(k)}\right) \tag{5.15}$$

In Equation (5.15) the kth user has a different binary data stream, so we label this $T_{p_n}^{(k)}$; however, to distinguish the user we must have a distinct time-hopping code, which we distinguish as $T_{c_n}^{(k)}$.

5.7.4 Orthogonal pulse multiple access system

As an example of OPM let us look closely at the modified Hermite orthogonal pulse system proposed by [17].

An M-ary communication system can be constructed from any set of orthogonal pulse shapes, such as $h_n(t)$ or $p_n(t)$. For simplicity, let us consider only modified Hermite pulse (MHP) waveforms.

We arbitrarily assume that 2-bit binary codes 00, 01, 10, and 11 are represented by MHP pulses of orders $n = 1, 2, 3, 4$. By assigning multiple-bit patterns to single pulse shapes, higher data rates can be achieved than simply by sending different pulse shapes. Furthermore, this can be extended to a coded scheme if desired.

Since MHP pulses are orthogonal, a multi-user system can be created using the same four pulse shapes, for example, by assigning MHP pulses of order $n = 1, 2$ to user 1 for the binary $0, 1$ and $n = 3, 4$ to user 2 for $0, 1$.

For a binary communication system using OPM, we wish to know whether a pulse representing either 0 or 1 is received. To achieve this, we need to generate a local copy of each pulse shape and integrate it with the received pulse. Conventionally, two complete sets of hardware are needed in order to produce two pulses of different shapes. However, because MHP shapes of lower order can be generated by integrating a pulse of higher order, a low-complexity multiple pulse generator can be constructed. We use the modified Hermite orthogonal pulse of the particular order from the first pulse generator to generate a different-order modified Hermite orthogonal pulse at the second pulse generator. The configuration of the second pulse generator is much less complex than if the different-order modified Hermite orthogonal pulses were produced from scratch based on a source signal. Also, only a single source signal is required to produce two different-order pulses.

A Matlab model of the circuit to obtain double pulse generation is outlined in Figure 5.9. We can see that by utilizing one of the properties of the pulses (i.e. by differentiating or integrating them) another pulse can be created, with the order of the pulses being one more or one less than the original pulse, respectively. In Figure 5.9 the order $n = 2$ is input to the system, along with a pulse of specified width. The width of the input pulse is determined by the desired width of the output pulse and is approximately twice the length of the output pulse. While the input pulse is on the pulse will be produced, but once the input pulse is zero any output will be suppressed. In an actual circuit the power would be removed from the input, so that no output would result in any case. In particular, the additional pulse is created by integrating the output from a pulse of order n. Thus, a pulse of order $n - 1$ is created.

5.8 CAPACITY OF UWB SYSTEMS

The capacity of a UWB multiple access system that is dependent on a specific pulse shape is computed by Zhao and Haimovich in [55], and, although this is a simplistic pulse shape and is only applicable for a PPM system, it serves to illustrate several points about UWB capacity. Thus, we follow the derivation here.

First, several assumptions are made. For simplicity and without any loss of generality, each UWB pulse is assumed to represent one symbol. Thus, the number of pulses per symbol $N_p = 1$. This means that the symbol interval time T_s and the frame time interval T_f are the same (i.e. $T_f = T_s$), and the energy per pulse E_p is the same as the energy per symbol.

Next, we comment that the Shannon capacity formula $C = W \log_2(1 + \text{SNR})$ applies to a channel with continuous valued inputs and outputs, which is not the case for PPM UWB where the values are discrete.

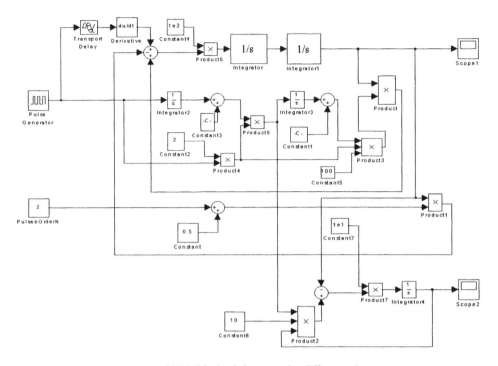

(a) Matlab circuit for generating different pulses.

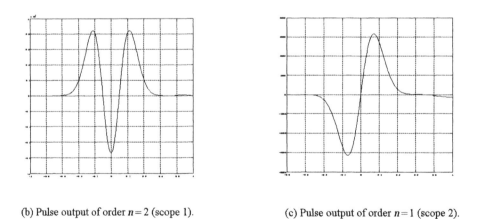

(b) Pulse output of order $n=2$ (scope 1).   (c) Pulse output of order $n=1$ (scope 2).

Fig. 5.9 (a) A circuit for generating multiple orthogonal pulses; (b) and (c) sample output pulses when the input is $n = 2$.

For simplicity, the pulse shape is a rectangular waveform $p(t)$ where

$$p(t) = \sqrt{\frac{1}{T_p}}, \quad 0 \leq t \leq T_p \tag{5.16}$$

and T_p is the pulse time.

The correlation function $h(\tau)$ for $p(t)$ is

$$h(\tau) = \begin{cases} \dfrac{T_p + \tau}{T_p}, & -T_p \leq \tau \leq 0 \\[2mm] \dfrac{T_p - \tau}{T_p}, & 0 \leq \tau \leq T_p \end{cases} \tag{5.17}$$

In this derivation we model the time shift difference between users, with the symbol Δ as a uniform distribution over the interval $[-T_f, T_f]$. It follows that the probabilities

$$P(-T_p \leq \Delta \leq 0) = P(0 \leq \Delta \leq T_p)$$

$$= \frac{T_p}{2T_f}$$

$$= \frac{1}{2\beta} \tag{5.18}$$

where $\beta = T_f/T_p$ is the spreading ratio, which is defined as the time of the frame divided by the time of the pulse.

The mean value of $h(\Delta)$ can be calculated as

$$E[h(\Delta)] = E[h(\Delta) \mid -T_p \leq \Delta \leq 0]P(-T_p \leq \Delta \leq 0)$$
$$+ E[h(\Delta) \mid 0 \leq \Delta \leq T_p]P(0 \leq \Delta \leq T_p) \tag{5.19}$$

$$= \frac{1}{2\beta} \tag{5.20}$$

The variance of $h(\Delta)$ is denoted as σ_h^2. Since $E[h(\Delta)] = 1/(2\beta)$, we can write

$$\sigma_h^2 = E[h^2(\Delta)] - \left(\frac{1}{2\beta}\right)^2 \tag{5.21}$$

$$= E[h^2(\Delta) \mid -T_p \leq \Delta \leq 0]P(-T_p \leq \Delta \leq 0)$$
$$+ E[h^2(\Delta) \mid 0 \leq \Delta \leq T_p]P(0 \leq \Delta \leq T_p) - \frac{1}{4\beta^2} \tag{5.22}$$

$$= \frac{1}{3\beta} - \frac{1}{4\beta^2} \tag{5.23}$$

which can be approximated as

$$\sigma_h^2 \approx \frac{1}{3\beta} \tag{5.24}$$

for large values of the spreading ratio β.

The variance σ_I^2 of the multiple access interference term N_I can be calculated as

$$\sigma_I^2 = \sum_{\nu=2}^{N_u} A^{(\nu)^2} \sigma_h^2 \approx \sum_{\nu=2}^{N_u} \frac{A^{(\nu)^2}}{3\beta} \tag{5.25}$$

where ν is the user's index and N_u denotes the total number of users.

The SNR at the output of the receiver for each symbol can be calculated as

$$\begin{aligned} \rho_I &= \frac{A^{(1)^2}}{\sigma_I^2 + N_0/2} \\ &= \frac{A^{(1)^2}}{\left(\sum_{\nu=2}^{N_u} A^{(\nu)^2}/3\beta\right) + N_0/2} \end{aligned} \tag{5.26}$$

after considering additive white Gaussian noise (AWGN).

With perfect power control, $A^{(1)} = A^{(\nu)} = \sqrt{E_p}$ and ρ_I can be expressed as

$$\rho_I = \frac{3\beta}{(N_u - 1) + 3\beta/\rho_0} \tag{5.27}$$

where $\rho_0 = 2E_p/N_0$. From this expression it can be seen that for a low number of users (i.e. $N_u < 3\beta/\rho_0$) the performance is noise-limited, while for a large number of users (i.e. $N_u > 3\beta/\rho_0$) the performance becomes interference-limited.

Using Equation (5.27) the single-user capacity $C_{M-\text{PPM}}$ as a function of the channel symbol SNR ρ_I is given by

$$C_{M-\text{PPM}}(\rho_I) = \log_2 M - E_{v|x_1} \log_2 \sum_{m=1}^{M} e^{\sqrt{\rho_I}(v_m - v_1)} \tag{5.28}$$

measured in bits per symbol. Here, v_m are random variables with m between 1 and M, and they have the following distribution conditional on the transmitted signal x_1:

$$\begin{aligned} v_1 &: N(\sqrt{\rho_I}, 1) \\ v_m &: N(0, 1), \quad m \neq 1 \end{aligned} \tag{5.29}$$

In Figure 5.10 the effects of thermal noise are ignored, and we concentrate on multiple access interference. The multiple access channel will achieve full user capacity when the number of users $N_u < 15$, and when the number of users is greater, user capacity will decrease.

5.9 COMPARISON OF UWB WITH OTHER WIDEBAND COMMUNICATION SYSTEMS

In this section we discuss some of the important differences and similarities of UWB communication systems, spread spectrum (SS), and OFDM systems. Although this

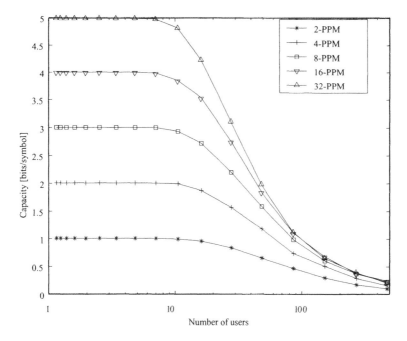

Fig. 5.10   User capacity for a multi-user UWB as a function of the number of users N_u for spreading ratio $\beta = 50$. Reproduced by permission of ©2002 IEEE.

is a general discussion, we have in mind indoor WLAN protocols (i.e. IEEE 802.11b and 802.11a). The reason is that in such a short space we cannot hope to cover in any depth the variations of SS communication or OFDM, and, more particularly, many of the applications in which SS or OFDM are used or planned to be used are not within the current sphere of interest of UWB. To give some examples, SS is used within the so-called *third-generation* mobile telephone and data services. The communication from base station to mobile is of the order of hundreds of meters to kilometers. On the other hand, OFDM is being considered for so-called *fourth-generation* mobile systems. Moreover, OFDM is also used for digital television broadcasting, such as ISDB-T in Japan. UWB communication techniques are not currently being considered for these outdoor, long-range applications. However, indoor WLANs are within the possible application sphere of UWB and, thus, present a good opportunity to compare and contrast.

The IEEE standards by which these WLANs are commonly known are 802.11b that employs a direct sequence spread spectrum (DSSS) centered at 2.4 GHz, and 802.11a that uses OFDM at 5 GHz. It will be helpful to review each of these in turn. Then, we will proceed to a comparison, and discuss the theoretical and practical difference between the three wideband communication systems.

5.9.1 CDMA

One of the most popular indoor wireless communication standards is the IEEE 802.11b standard for WLANs. It operates in the 2.4 GHz unlicensed band. In 802.11b, SS techniques are used to take a narrowband data signal and spread it over the entire available frequency band, in order to combat interference from other users or noise sources. The 2.4 GHz band is known as the ISM band, which stands for industrial, scientific, and medical band. It accommodates many sources of electromagnetic radiation. One of the most common of these is the ordinary microwave oven.

There are two common techniques to spread the spectrum: the frequency-hopping spread spectrum (FHSS) and the DSSS. An overview of the frequency–time relationship is shown for these two methods in Figures 5.11 and 5.12.

In Figure 5.11 we can see that two users occupy a narrow frequency band for a short period of time. There are 79 frequency-hopping channels in the IEEE 802.11 standard and each is 1 MHz in width. Hops must take at most 224 µs.

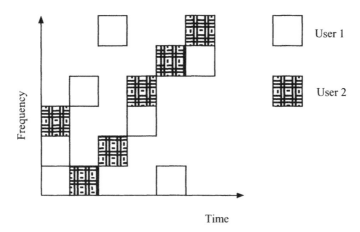

Fig. 5.11   Frequency–time relationship for two users using the FHSS.

In contrast, Figure 5.12 shows that each user occupies all of the available spectrum all the time and that different users are separated by their PN codes. Thus, DSSS is also called CDMA.

Although both DSSS and FHSS are specified as standards for IEEE 802.11 WLAN, in the more recent IEEE 802.11b standard DSSS is the only physical layer defined. The 802.11a standard defines an OFDM physical layer (discussed later).

5.9.2   Comparison of UWB with DSSS and FHSS

In [56] a comparison of three different modulation techniques was considered: DSSS, FHSS, and UWB. The setting was for each method to occupy 3.2 MHz, transmit at a rate of 3.125 Mbps, and support 30 simultaneous users.

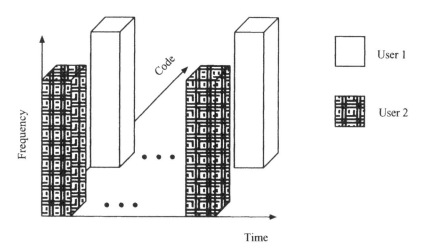

Fig. 5.12 Frequency–time relationship for two users using the DSSS. The two users are separated by different codes.

For a DSSS system the SNR can be expressed as

$$\text{SNR}_{\text{DSSS}} = \frac{1}{(K-1)/3N - N_0/2E_b} \tag{5.30}$$

where K is the number of users and N is the number of chips per bit. The BER can then be calculated from

$$\text{BER}_{\text{DSSS}} = \frac{1}{2}\,\text{erfc}\left(\sqrt{\frac{\text{SNR}_{\text{DSSS}}}{2}}\right) \tag{5.31}$$

where 'erfc' is the complementary error function, which is defined as

$$\text{erfc}(z) = 1 - \frac{1}{\sqrt{\pi}}\sum_{k=0}^{\infty}\frac{(-1)^k z^{2k+1}}{k!(2k+1)} \tag{5.32}$$

On the other hand, the BER for FHSS can be calculated as

$$\begin{aligned}
\text{BER}_{\text{FHSS}} = &\frac{1}{2}\left(1 - \frac{1}{k}\right)^{M-1}\text{erfc}\left(\sqrt{\frac{S}{2N}}\right) \\
&+ \frac{1}{2}\sum_{i=1}^{M-1}\left(\frac{1}{k}\right)^{i}_{M-1}C_i\,\text{erfc}\left(\sqrt{\frac{S}{2N+S_i}}\right)
\end{aligned} \tag{5.33}$$

where k is the number of frequency-hopping slots, M is the number of users, S is the signal power, and N is the noise power. Thus, S_i represents the signal power from interfering users.

For the UWB result, the average output SNR can be calculated by assuming a random time-hopping sequence. Let the number of active users be N_u. From [10] the SNR is

$$\text{SNR} = \frac{(N_s A_1 m_p)^2}{\sigma_{\text{rec}}^2 + N_s \sigma_a^2 \sum_{k=2}^{N_u} A_k^2} \tag{5.34}$$

where σ_{rec}^2 is the variance of the receiver noise component at the pulse train integrator output. The monocycle waveform-dependent parameters m_p and σ_a^2 are given by

$$m_p = \int_{-\infty}^{\infty} \omega_{\text{rec}}(x - \delta) v(x) \, dx \tag{5.35}$$

and

$$\sigma_a^2 = T_f^{-1} \int_{-\infty}^{\infty} \left[\int_{-\infty}^{\infty} \omega_{\text{rec}}(x - s) v(x) \, dx \right]^2 ds \tag{5.36}$$

respectively [56], where A_1 is the monocycle amplitude, T_f is the frame time which is assumed to be 10 ns, and N_s is the number of impulses per symbol.

The first result is shown in Figure 5.13 for the single-user case where we can easily see that all three methods have the same BER curve against SNR.

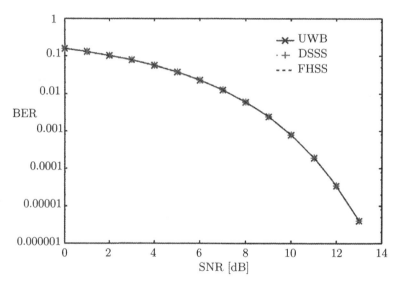

Fig. 5.13 Comparison of the BER of three wideband systems DSSS, FHSS, and UWB for a single user.

When 30 users are transmitting simultaneously, differences are seen. This is illustrated in Figure 5.14. An error floor is seen for frequency hopping, which is because the number of users is too large for the number of frequency slots available. Thus, collisions will always occur, generating interference even at high SNR.

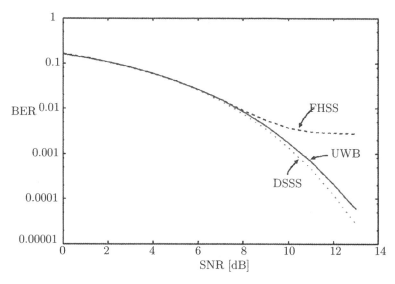

Fig. 5.14 Comparison of BER for the three systems when 30 users are simultaneously transmitting.

An expanded view of Figure 5.14 is shown in Figure 5.15, which clearly shows that, theoretically under these assumed conditions, DSSSdirect sequence spread spectrum (DSSS) performs better than UWB. However, the chip bandwidth assumed for DSSS is 0.37 ns, which means that signal processing is much more difficult (and therefore more expensive) to produce than for UWB systems.

Thus, we can summarize by saying that similar performance is obtained by both DSSS and UWB systems, given the same bandwidth constraints. However, from the practical perspective, UWB offers a possibly much cheaper implementation. As the bandwidth increases, the signal-processing burden on DSSS and FHSS systems increases, making UWB more attractive. We further note that the bandwidth assumed in this example is only 3 MHz: UWB offers bandwidth possibilities that are at least three orders of magnitude greater.

5.9.3 OFDM

With OFDM, multiple orthogonal carriers are transmitted simultaneously. By transmitting several symbols in parallel, symbol duration is increased proportionately, which reduces the effects of ISI caused by a dispersive Rayleigh-fading environment.

The input sequence that determine which of the carriers is transmitted during the signaling interval is

$$s(t) = Ae^{2\pi f_i t} \cdot \Pi(t/T) \qquad (5.37)$$

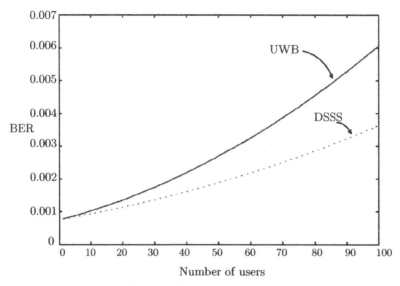

Fig. 5.15 Comparison of BER against the number of users for UWB and DSSS systems.

where

$$f_i = f_c + \frac{i}{T}, \quad i = 0, 1, \ldots, N - 1 \tag{5.38}$$

and

$$\Pi(t/T) = \begin{cases} 1, & -\dfrac{T}{2} \leq t \leq \dfrac{T}{2} \\ 0, & \text{otherwise} \end{cases} \tag{5.39}$$

The total number of sub-band carriers is N, and T is the symbol duration for the information sequence. In order that the carriers do not interfere with each other, the spectral peak of each sub-carrier must coincide with the zero crossings of all the other carriers (i.e. the sub-carriers are orthogonal). This is shown graphically in Figure 5.16. The difference between the center lobe and the first zero crossing represents the minimum required spacing and is equal to $1/T$. An OFDM signal can be constructed by assigning parallel bitstreams to the sub-band carriers.

A block diagram of a typical 802.11a OFDM transmitter is shown in Figure 5.17. The modulation symbols are mapped to the sub-carrier of the 64-point inverse discrete Fourier transform (IDFT) to create an OFDM symbol. However, only 48 sub-carriers are used for data modulation, four sub-carriers are used for pilot tones (which are used for channel estimation), and 12 sub-carriers are not used.

The output of the inverse fast Fourier transform (IFFT) is converted to a serial sequence, and a guard interval or cyclic prefix is added.

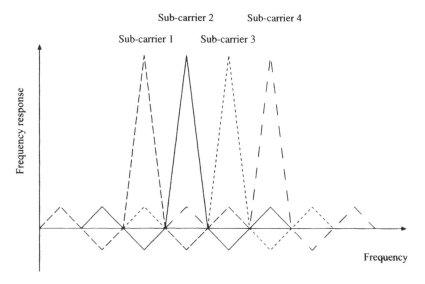

Fig. 5.16 Graphical representation of four orthogonal sub-carriers to make up an OFDM symbol.

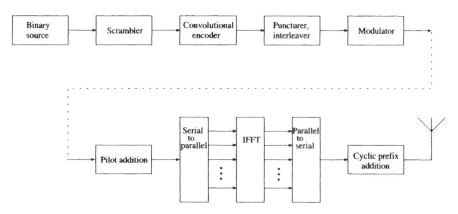

Fig. 5.17 Block diagram of a typical OFDM transmitter (IEEE 802.11a standard).

A block diagram of a typical 802.11a OFDM receiver is shown in Figure 5.18. The receiver performs the inverse of the operations of the transmitter. Table 5.2 summarizes the key parameters of the IEEE 802.11a OFDM WLAN standard.

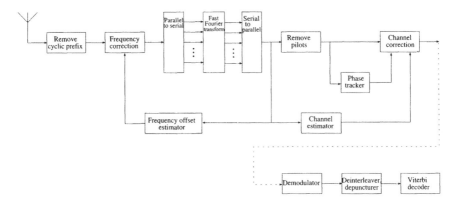

Fig. 5.18 Block diagram of a typical OFDM receiver (IEEE 802.11a standard).

Table 5.2 Key parameters of IEEE 802.11a OFDM WLAN standard (QPSK, quadrature phase shift keying; QAM, quadrature amplitude modulation).

Parameter	Values
Data rate	6, 9, 12, 18, 24, 36, 48, 54 Mbps
Modulation	BPSK, QPSK, 16-QAM, 64-QAM
Sub-carriers	52
Pilot tones	4
Symbol duration	4 μs
Sub-carrier spacing	312.5 kHz
Signal bandwidth	16.66 MHz
Channel spacing	20 MHz

5.10 INTERFERENCE AND COEXISTENCE OF UWB WITH OTHER SYSTEMS

Since UWB signals have such a broad bandwidth, they operate as an *overlay system* with other wireless communication methods. Figure 5.19 shows some of the wireless systems with which UWB must contend. The problem is twofold. First, UWB must mitigate or be able to operate in the presence of these interferers. Second, UWB must not provide substantial interference to users of these other services. The definition of what is substantial interference is varied, but is generally taken to mean less than the legal maximum or less than the interference from unintentional radiators of electromagnetic energy.

We can easily see that one of the reasons for avoiding the lower frequency bands is the plethora of wireless services with which UWB must contend. Between 3 GHz and 10 GHz the main source of interference for indoor wireless systems is assumed to be the 5 GHz WLANs which are based on OFDM.

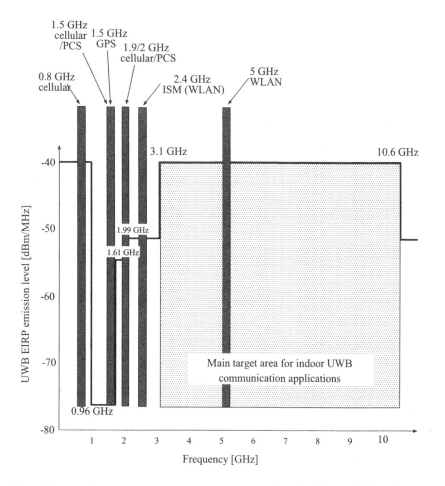

Fig. 5.19 Other wireless systems operating in the same bandwidth as UWB will both cause interference to and receive interference from each other (ISM, industrial, scientific and medical).

5.10.1 WLANs

In [57] the interference from IEEE 802.11a WLANs on UWB systems is discussed. It concluded that, with the interfering system center frequency at 5.2 GHz and the UWB system center frequency at 4.2 GHz, the effect on BER performance is minimal when the signal power of the 802.11a signal is less than 10 dB of the UWB signal. The UWB system was simulated using a pulse waveform $s_0(t)$, which was modulated by a sine wave $\sin(2\pi f_0 t)$ at frequency f_0 (in hertz) and given by the equation

$$s_0(t) = e^{-at^2/\tau^2} \sin(2\pi f_0 t) \tag{5.40}$$

where, in the simulation, $a = \ln 10$ and $\tau = 0.5$ ns. Pulses were sent every 5 ns using BPM. The power of the UWB system for E_b/N_0 (energy per bit divided by the noise power density) and the desired-to-undesired (DU) ratio was defined as the power of a single pulse in order to remove the effect of the pulse interval.

When the 802.11a signal is stronger, such as when the WLAN transmitter is close to the UWB receiver, significant interference occurs. To mitigate this interference, two techniques are suggested. One is to use a filter to remove unwanted interference. For example, a sixth-order Chebyshev filter with cutoff frequency at 4 GHz, ripple less than 0.2 dB, and -20 dB attenuation at 5.18 GHz can be used. This filter causes a 1 dB loss when there is no interferer present; however, significant performance improvement (no error floor for a DU signal ratio of 0 dB) is obtained when the interfering signal is present.

The second proposal is to use a multiband UWB system. The highest frequency sub-carriers are removed because they overlap with the 802.11a spectrum. Using a system with 11 sub-carrier pulses at intervals of 200 MHz from 3.2 to 5.2 GHz gives

$$s_0(t) = e^{-at^2/\tau^2} \sin(2\pi f_0 t) \tag{5.41}$$

$$\approx \sum_{n=1}^{11} X_n e^{-at^2/\tau'^2} \sin(2\pi f_0 t) \tag{5.42}$$

where $\tau' = 0.5$ ns.

It was concluded that, at a DU ratio of 0 dB, removal of the two highest sub-carriers gave the best BER performance, while, at a DU ratio of -10 dB, removal of three sub-carriers gave the best performance.

An experiment was performed to measure the performance of 802.11b WLAN networks under the influence of high-power UWB signals, and the results were reported in [58]. It was concluded that 802.11b can suffer from UWB interference given high-powered transmission and a close range between the UWB transmitter and the WLAN receiver. The experimental setup is shown in Figure 5.20.

In this experiment, pulses with width 500 ps were generated from several UWB transmitters. The PRF in these prototypes was fixed at 87 MHz. The center frequency of the transmission was around 1.8 GHz. The peak-to-peak voltage for the pulse measured from the output port of the circuit board was approximately 300 mV. Omni-directional antennas were employed and had EIRP of -2 to 3 dBm depending on the PRF. These prototypes are not compatible with FCC regulations, exceeding the mask for indoor transmission by 30 dB in the 2.4 GHz band.

Spectrum measurements on channel 1 ($f_c = 2.412$ GHz) showed peak UWB interference approximately 20 dBmV less than the peak 802.11b signal with 20 UWB transmitters at 100 cm. At 15 cm from the measuring antenna the UWB peak was approximately 5 dBmV less than the peak WLAN signal.

Signal-to-noise measurements were undertaken, and the conclusion was reached that, with the UWB transmitters used, if the distance from the WLAN receiver is greater than 50 cm, then no significant reduction in SNR occurred. For distances less than 50 cm, the SNR reduction was as much as 10 to 15 dB.

Fig. 5.20 Experimental setup used to find the interference to a wireless LAN card from high-powered UWB transmitters. Reproduced by permission of ©2003 IEEE.

Throughput measurements showed a similar trend: with distances less than 30 cm, WLAN throughput dropped dramatically when 15 or more high-powered UWB transmitters were used and if the distance was greater than 40 cm the deterioration was negligible.

5.10.2 Bluetooth

In [58] the performance of Bluetooth WLANs was considered under the influence of high-powered UWB signals.

It was concluded that the Bluetooth connection suffered less than the corresponding 802.11b channel. The reason for this was that the fixed PRF was used for the UWB transmitters, which gave a distinct line spectrum. Since Bluetooth devices are able to monitor individual channel states these 'bad' channels can be avoided. Only a degradation in throughput from 530 kbps to approximately 490 kbps was recorded for Bluetooth transceivers separated by 10 m, and no degradation in throughput was recorded for Bluetooth devices separated by 3 m. In these measurements the UWB devices were located in an arc 15 cm from the Bluetooth receiver.

5.10.3 GPS

In [59] a report into the interference of UWB transmitters on both high-grade aviation GPS receivers and also low-cost OEM (original equipment manufacturer) GPS receivers was made.

The report concluded that UWB transmitters could indeed affect GPS receivers and that care must be taken for transmission at GPS frequencies. Based on this report and others, the FCC mandated a strict limit on emissions within the GPS frequency band.

In 1999 the US Department of Transportation (DOT) approached Stanford University to research the compatibility of UWB and GPS and to conduct tests to help quantify any interference problems. GPS plays a major role in many commercial, military, and public systems. For example, aircraft rely on the information from GPS, as do most modern vehicles equipped with navigation systems. Mobile phones equipped with GPS receivers have been commercialized. It is anticipated that reliance on GPS and similar satellite systems will increase in the future.

The majority of the tests performed by Stanford University measured the UWB impact on the accuracy and loss-of-lock performance of a high-grade GPS aviation receiver. A smaller test set measured the UWB impact on the loss-of-lock performance for two different receivers: an aviation receiver and a low-cost commercial receiver. Tests were also undertaken to measure the UWB impact on the signal acquisition performance of a high-grade, general-purpose GPS receiver. This third receiver used the same hardware as the aviation receiver, but the firmware was changed so that the receiver did not utilize an acquisition strategy suited for aircraft dynamics.

It was found that spectral lines present in any UWB signal degraded the GPS signal significantly. A 17 dB degradation was measured without making any effort to place the UWB signals on the more sensitive GPS spectral lines. In practice, UWB lines will frequently find more sensitive lines than those found in these trials because (a) many GPS satellites will be in view and (b) the Doppler frequency for each satellite will change as the satellite moves across the sky, causing the frequency of the more sensitive lines to shift. Eventually, sensitive lines from one satellite or another will fall on the spectral lines from any nearby UWB transmitter that has such lines. The worst line for GPS satellite PRN 21 is 6.5 dB more sensitive than the victim line in these measurements.

Under the best circumstances, UWB signals with high PRFs appear as broadband noise. In other words, the equivalence factor is approximately 0 dB but only in the absence of in-band spectral lines. If UWB dithering codes or modulation indices are not chosen carefully and some spectral line content remains, then the UWB waveform is more damaging than white noise.

It was found that trends were observed in all three receivers: aviation receiver, general-purpose receiver, and commercial receiver.

All tests showed the same sensitivity to UWB signal type. For example, the worst interference cases for all three receivers occurred when a discrete UWB spectral line fell into the GPS band. However, OEM tests must be more carefully interpreted because the OEM front end bandwidth is significantly narrower than the bandwidth for the aviation receiver and the standard filter used to measure noise power.

5.10.4   Cellular systems

In comparison with the GPS and other indoor communications techniques the effects of wireless UWB systems on cellular mobile telephone services has not been well covered in the literature. This is probably not surprising, recalling Figure 5.19. Most current mobile telephone systems fall below 1 GHz and, thus, are not in the frequency band in which it is anticipated that most UWB communication systems will use. The 1.5 GHz cellular band is heavily protected by current FCC regulations. It can be anticipated that 2 GHz services will be somewhat affected; however, this is at the edge of the main UWB bandwidth and is not as popular as other cellular services at the present time.

Some experiments have been performed. For example, in [60] the effect of a prototype UWB communication system meeting FCC requirements on a 1.9 GHz PCS band mobile phone was examined. Although the tests considered subjective voice quality only, no discernible difference was found by the people making the telephone calls. In these tests the mobile phone was located approximately 1.5 m away from the UWB transmit antenna.

5.10.5   Wi-Max

In the modern world many people feel the need for fast Internet connection. In the literature, broadband transmission generally refers to transmission rates of greater than 1.5 Mbps. Cable modem systems, digital subscriber lines (DSL) and T1 connections are among the means of providing consumers with fast wired connectivity to the Internet. All of these wired systems require a reliable infrastructure, not only within the home but also in the backbone connections using material such as coaxial cable and fiber optic cable for the connection. However, in rural areas and developing countries such installation of cable simply does not exist and Internet service providers do not see desirable enough profits to go through the hassle of cable installations. This problem also exists in urban areas which lack proper infrastructure. This means that a great number of users will have to do without broadband access.

A wireless solution can address this problem. Wireless systems have the capacity to address broad geographic areas and the implementation cost for these systems is much less than wired-based systems. Wi-Max (world interoperability for microwave access) is a technology that aims to provide wireless last-mile broadband access with fast connection speeds comparable to T1, cable, and DSL connections. Wi-Max will provide connectivity with speeds up to 75 Mbps over ranges up to 30 miles. It is believed that, in its first implementations, Wi-Max will bring the network to a building, while users inside the building will be able to get connected using in-building technologies such as WLAN or Ethernet.

There are features and techniques used in the development of Wi-Max that make this technology able to fulfill its promise. Adaptive modulation is among these techniques. Different order modulations allow the transmitter to send more bits per symbol which in turn will lead to higher throughput and better spectral efficiencies. It must be noted that when using higher rate modulation techniques such as 64 QAM, a better SNR is needed to maintain a specific BER. Adaptive modulation is one technique which will enable Wi-Max compatible systems to choose the highest order modulation possible for the existing channel conditions.

The working group responsible for developing the standard for Wi-Max is IEEE 802.16. The group initially developed the standard for the 10 to 66 GHz part of the spectrum. Very short wavelengths along with the problems of communication in highly cluttered environments, made this mode suitable for LOS communications. LOS requires the antennas to be mounted on heights that might not be suitable for urban environments. This shifted the focus to the 2 to 11 GHz portion of the spectrum which would make NLOS communications an option for Wi-Max. This led to the IEEE 802.16a amendment which discussed the NLOS mode of IEEE 802.16. The final version of the standard was released in 2004. The release of this version, which is referred to as IEEE 802.16-2004, was the completion of the essential fixed wireless standard behind Wi-Max. The IEEE 802.16 working group is now working on the IEEE 802.16e amendment which aims to add mobility to Wi-Max. Table 5.3 demonstrates the five different modes for fixed broadband wireless access (FBWA), specified in IEEE 802.16-2004.

Table 5.3 Five different modes for FBWA, specified in IEEE 802.16-2004.

Designation	Applicability
WirelessMAN-SC	10–66 GHz
WirelessMAN-SCa	Sub 11 GHz licensed bands
WirelessMAN-OFDM	Sub 11 GHz licensed and license-exempt bands
WirelessMAN-OFDMA	Sub 11 GHz licensed bands
WirelessHUMAN	Sub 11 GHz license-exempt bands

5.10.5.1 Effect of UWB interference on Wi-Max systems The coexistence issue of UWB systems with narrowband systems has been investigated in the literature. However, the effect of UWB interference on Wi-Max systems has not been significantly addressed at the present time. This issue has been analyzed in [61] by comparing the interference produced by a UWB hot spot with threshold interference values for different modes of operation defined for Wi-Max systems.

Results were generated by varying different parameters including victim receiver bandwidth, carrier frequency, activity factor, number of UWB devices, and the minimum distance that UWB devices can be from the victim receiver.

By investigating through simulation the interference caused by a realistic UWB hot spot located in the vicinity of a Wi-Max/IEEE 802.16 based receiver, it was concluded that in a realistic hot spot scenario (with a reasonable number of UWB devices that are active for a reasonable amount of time, assuming a uniform distribution), UWB devices pose no threat to the operation of IEEE 802.16 systems and the aggregate interference caused by them will fall below the threshold specified in the standard. It should be noted, however, that increasing the number of UWB devices or placing the majority of devices close to the victim receiver might damage the operation of the receiver.

It was also concluded that having high activity factors would increase the chances of having harmful interference. These problems could be tackled by developing some form of power control for the UWB system. It can be concluded that, of the IEEE 802.16 modes, the only mode that might have problems in the vicinity of a UWB hot spot is WirelessMAN-SCa when operating in low bandwidths, as it has a threshold much lower than the other proposed modes. The analysis has shown, however, that in realistic and practical cases a Wi-Max/IEEE 802.16-based system can operate in all modes without any significant problems in the presence of a UWB hot spot.

5.10.6 The effect of narrowband interference on UWB systems

UWB impulse radio has inherent immunity to narrowband interference (NBI), because of the very short time windowing used at the correlator receiver. However, due to the very low PSD allowed for UWB communications, it is expected that even this inherent NBI immunity is not sufficient to resist high levels of NBI; for example, an IEEE 802.11a WLAN device operating nearby would be seen as NBI by UWB receivers. The UWB receiver could be jammed if such interference is not properly suppressed.

At the output of a rake-type coherent receiver, the decision statistic is usually given by $\mathbf{c}^H \mathbf{r}$, in which $\mathbf{c}$ is the weighting vector used for combining and $\mathbf{r}$ is the vector that collects rake output samples from different paths over a symbol interval. In the presence of NBI and AWGN, the rake output vector $\mathbf{r}$ can be expressed as $\mathbf{r} = \mathbf{s} + \mathbf{i} + \mathbf{n}$, where $\mathbf{s}$ is the vector containing the desired signal samples, and $\mathbf{i}$ and $\mathbf{n}$ contain rake output samples from the NBI and AWGN, respectively. Here, all $\mathbf{r}$, $\mathbf{s}$, $\mathbf{i}$, $\mathbf{n}$, and $\mathbf{c}$ are $N_r \times 1$ vectors, with N_r denoting the number of resolvable multipath components.

A conventional rake receiver employs the weight vector $\mathbf{c} = \mathbf{h}$ to perform maximum ratio combining (MRC), where $\mathbf{h}$ is the vector consisting of channel fading coefficients. MRC maximizes the receiver's output SNR when only AWGN exists. However, in the presence of NBI, the interference samples at the rake output are correlated and the choice of $\mathbf{c} = \mathbf{h}$ is no longer optimal. Under the assumption that the UWB signal, NBI, and AWGN are mutually independent, the receiver output signal-to-interference plus noise ratio (SINR) is given by

$$\mathrm{SINR}_{\mathrm{MRC}} = \frac{N_s E_s \left| \mathbf{h}^H \mathbf{h} \right|^2}{\mathbf{h}^H \left(\mathbf{R}_i + \mathbf{R}_n \right) \mathbf{h}} \tag{5.43}$$

where E_s is the signal energy per symbol, N_s is the number of pulses transmitted per symbol, and

$$\mathbf{R}_n = E\left\{\mathbf{n}\,\mathbf{n}^{\mathrm{H}}\right\} \tag{5.44}$$

where superscript H denotes the conjugate transpose operation and

$$\mathbf{R}_i = E\{\mathbf{i}\,\mathbf{i}^{\mathrm{H}}\} \tag{5.45}$$

are the correlation matrices of the noise and interference samples collected over a symbol interval at different paths, respectively. It has been shown [62] that each element of $\mathbf{R}_i$ is proportional to the spectral power of the received UWB pulse at the NBI center frequency f_i, i.e. $|W(f_i)|^2$, where

$$W(f) = \int_{-\infty}^{\infty} w(t)e^{-j2\pi ft}dt \tag{5.46}$$

and $w(t)$ is the received UWB pulse waveform. This indicates that the impact of NBI would be the severest when the NBI center frequency overlaps with the nominal center frequency of the UWB spectrum.

In the presence of NBI, the optimal weighing vector $\mathbf{c}$ for rake combining is the one that minimizes the mean square error. The minimum-mean-square-error combining (MMSEC) vector has the form of

$$\mathbf{c} = \frac{N_s E_s \left(\mathbf{R}_i + \mathbf{R}_n\right)^{-1}\mathbf{h}}{1 + N_s E_s \mathbf{h}^{\mathrm{H}} \left(\mathbf{R}_i + \mathbf{R}_n\right)^{-1}\mathbf{h}} \tag{5.47}$$

The corresponding receiver output SINR is given by

$$\mathrm{SINR}_{\mathrm{MMSEC}} = N_s E_s \mathbf{h}^{\mathrm{H}} \left(\mathbf{R}_i + \mathbf{R}_n\right)^{-1}\mathbf{h} \tag{5.48}$$

Another NBI suppression method is frequency domain notch filtering. That is, the part of the UWB power spectrum $|W(f)|^2$ that is around the interference center frequency f_i is notched out. The transfer function and impulse response of the notch filter are defined, respectively, as

$$H(f) = 1 - \left[u\left(f - f_i + \frac{B}{2}\right) - u\left(f - f_i - \frac{B}{2}\right)\right] \tag{5.49}$$

$$h(t) = \delta(t) - Be^{j2\pi f_i t}\mathrm{sinc}(Bt) \tag{5.50}$$

where $u(\cdot)$ is the unit step function, $\delta(\cdot)$ is the Dirac delta function, and the spectral notch width B should be sufficiently larger than the bandwidth of the interference. In practice this could be implemented with a filter bank. The UWB pulse waveform at the output of the notch filter can be expressed as

$$w_{\mathrm{notch}}(t) = w(t) - W(f_i)Be^{j2\pi f_i t}\mathrm{sinc}(Bt) \tag{5.51}$$

where t is in nanoseconds and B is in gigahertz; thus, $B \ll 1$ for the suppression of typical NBI. Due to the low power of UWB signals, $W(f_i)$ is also of a very small value. Therefore, $w_{\mathrm{notch}}(t)$ is expected to be only slightly different from $w(t)$. However, the frequency domain notching needs to estimate the interference frequency and in practice, complete annulment will not be possible because of nonideal filters.

Figure 5.21 depicts the receiver output SINR as a function of the interference center frequency f_i for MRC, MMSEC, and frequency domain notch filtering, when the NBI is modeled as a band-limited signal with a constant power spectrum over a bandwidth of 20 MHz. The interference-to-signal ratio (ISR) per pulse is 30 dB and the spectral notch width is fixed at 40 MHz. The simulation results assume that the received pulse shape of the UWB signal is a Gaussian waveform with $t = 0.0678$ ns, the SNR per pulse is 0 dB, the data rate is 1 Mbps, and the pulse repetition time is long enough to guarantee no inter-frame interference. The MRC curve shows that the performance degradation is the severest when the nominal center frequencies for the NBI and UWB spectra are overlapping.

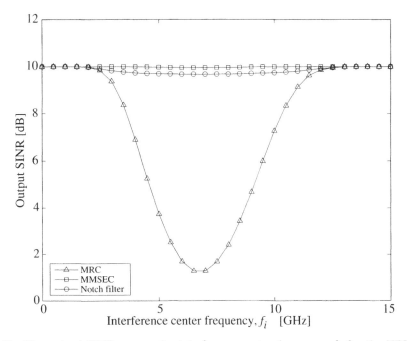

Fig. 5.21 The output SINR versus the interference center frequency f_i for the NBI with a constant power spectrum over a bandwidth of 20 MHz.

Figure 5.22 shows the BER for MRC, MMSEC, and frequency domain notch filtering, when the interference is from an OFDM-based WLAN device. The OFDM signal in this example has 52 sub-carriers with a total bandwidth of 16.56 MHz and a center frequency at 5.2 GHz, and the transmission from the interferer is assumed to be continuous with an ISR per pulse of 30 dB. The MRC curve again shows that significant performance loss occurs when no NBI suppression technique is applied. It can be clearly seen that both MMSEC and frequency domain notch filtering achieve effective NBI suppression.

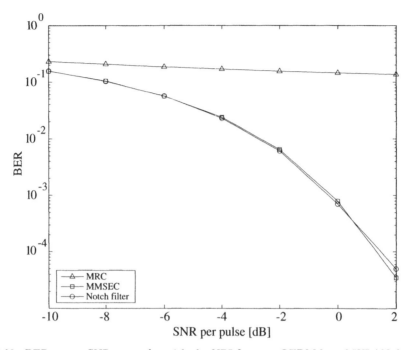

Fig. 5.22 BER versus SNR per pulse with the NBI from an OFDM-based WLAN device.

5.11   SUMMARY

In this chapter we examined UWB wireless communications. Modulation methods including PPM, BPM, OPM, and their combinations were presented. Receiver design and pulse detection were examined. Multiple access techniques for UWB communications were also examined.

A simple derivation of the capacity of the wireless UWB channel was presented. The effect of UWB on existing wireless communication methods, such as the IEEE 802.11 WLAN standards and Bluetooth, was examined. Several methods to prevent interference from these narrowband systems on UWB, in particular the use of filters and multiband UWB, were studied.

Finally, a brief discussion on the relative merits and demerits of UWB as a communication method with other wideband communication techniques, such as CDMA and OFDM, was undertaken. The effect of NBI was also examined.

Problems

Problem 1. Investigate PN codes and summarize three different codes. What is the processing gain of each PN code? Apply each of the codes to:

(a) an unmodulated pulse train;

(b) a 2-PPM pulse train; and

(c) a 4-PPM pulse train.

Compare the spectra obtained with 100 pulses and comment.

Problem 2. Implement the Matlab circuit-generating multiple orthogonal pulse shapes of Figure 5.9. Discuss the potential problems that might be present in practice.

Problem 3. In Figure 5.10 the capacity versus number of users is shown for a spreading ratio of $\beta = 50$. Calculate the curves for $\beta = 5$ and $\beta = 500$. At what number of users does the capacity begin to decrease? You may assume a pulse width of $T_p = 1$ ns and the transmission symbol rate $R = 1/(\beta T_p)$.

Problem 4. Calculate the aggregate capacity C of all traffic on the multiple access channel for all the PPM types shown in Figure 5.10. The aggregate capacity can be calculated from $C = N_u C_{M-PPM}$.

6

Advanced UWB pulse generation

The system of sine and cosine as orthogonal functions has been historically synonymous with communications. Whenever the term frequency is used, reference is made implicitly to these functions, and, hence, the general theory of communication is based on the system of sine and cosine functions.

In recent years other complete systems of orthogonal functions have been used for theoretical investigations as well as for equipment design. Analogs to Fourier series, Fourier transform, frequency, power spectra, and amplitude, phase, and frequency modulation exist for many systems of orthogonal functions. This implies that theories of communication can be worked out on the basis of these systems. Most of these theories are of academic interest only. In this chapter we will examine in detail how to generate pulse waveforms for UWB systems, for more complex orthogonal pulses.

6.1 HERMITE PULSES

For the complete system of the orthogonal Hermite functions, the implementation of circuits by modern semiconductor techniques appears to be competitive in a number of applications with the implementation of circuits for the system of sine and cosine functions. In this section we present Hermite pulses as one example of orthogonal pulses that can be used for UWB communications.

Hermite functions and, indeed, Hermite pulses are not new [63]. The Hermite transform has already been used to shed light on spatiotemporal relationships in image processing [64]. The use of modified Hermite polynomials to create an orthogonal transform and even to create orthogonal wavelets for multicarrier data transmission over high-rate digital subscriber loops has been proposed [65]. The application of these pulses to communications seems limited to the latter reference, and, with the exception of the similarity to Hermite polynomials mentioned in [38], there seems to have been no attempt to systematically apply modified Hermite pulses to UWB communications before the proposal in [17]. Here, we present a systematic analysis of Hermite pulses and modify them for use in UWB communication systems.

6.1.1 Hermite polynomials

Charles Hermite was born in 1822 in France. He was a very brilliant mathematician. Some of his mathematical ideas are still widely used today, especially the Hermitian forms that are used in physics and mathematics. Some of Hermite's other noted accomplishments that bear his name are Hermite polynomials, Hermite's differential equation, Hermite's formula of interpolation, and Hermitian matrices.

A polynomial is a finite sum of terms like $a_k x^k$, where k is a positive integer or zero. The functions defined by

$$h_{e_n}(t) = (-\tau)^n e^{t^2/2\tau^2} \frac{d^n}{dt^n} \left(e^{-t^2/2\tau^2} \right) \tag{6.1}$$

where $n = 0, 1, 2, \ldots$ and $-\infty < t < \infty$, are called *Hermite polynomials*. The parameter τ is the time-scaling factor. It should be mentioned that the definition of Equation (6.1) is one of various forms of Hermite polynomials used in the literature.

As examples of these polynomials we can write

$$h_{e_0}(t) = 1$$

$$h_{e_1}(t) = \frac{t}{\tau}$$

$$h_{e_2}(t) = \left(\frac{t}{\tau}\right)^2 - 1$$

$$h_{e_3}(t) = \left(\frac{t}{\tau}\right)^3 - 3\frac{t}{\tau}$$

$$h_{e_4}(t) = \left(\frac{t}{\tau}\right)^4 - 6\left(\frac{t}{\tau}\right)^2 + 3 \tag{6.2}$$

$$h_{e_5}(t) = \left(\frac{t}{\tau}\right)^5 - 10\left(\frac{t}{\tau}\right)^3 + 15\frac{t}{\tau}$$

$$h_{e_6}(t) = \left(\frac{t}{\tau}\right)^6 - 15\left(\frac{t}{\tau}\right)^4 + 45\left(\frac{t}{\tau}\right)^2 - 15$$

$$h_{e_7}(t) = \left(\frac{t}{\tau}\right)^7 - 21\left(\frac{t}{\tau}\right)^5 + 105\left(\frac{t}{\tau}\right)^3 - 105\frac{t}{\tau}$$

$$h_{e_8}(t) = \left(\frac{t}{\tau}\right)^8 - 28\left(\frac{t}{\tau}\right)^6 + 210\left(\frac{t}{\tau}\right)^4 - 420\left(\frac{t}{\tau}\right)^2 + 105$$

which are related by the following equations

$$h_{e_{n+1}}(t) = \frac{t}{\tau} h_{e_n}(t) - \tau \dot{h}_{e_n}(t) \tag{6.3}$$

$$\dot{h}_{e_n}(t) = \frac{n}{\tau} h_{e_{n-1}}(t) \tag{6.4}$$

where ' ' ' stands for derivative with respect to time.

Using Equations (6.3) and (6.4) the differential equation which is satisfied by Hermite polynomials is derived as

$$\tau^2 \ddot{h}_{e_n}(t) - t \dot{h}_{e_n}(t) + n h_{e_n}(t) = 0 \tag{6.5}$$

6.1.2 Orthogonal modified Hermite pulses

By definition, two real-valued functions $g_m(t)$ and $g_n(t)$ that are defined on an interval $a \le x \le b$ are orthogonal if

$$(g_m \cdot g_n) \doteq \int_a^b g_m(t) g_n(t)\, dt = 0, \quad m \ne n \tag{6.6}$$

A set of real-valued functions $g_1(t), g_2(t), g_3(t), \ldots$ is called an orthogonal set of functions in the set. The nonnegative square root of $(g_m \cdot g_m)$ is called the norm of $g_m(t)$ and is denoted by $\|g_m\|$; thus,

$$\|g_m\| = \sqrt{(g_m \cdot g_m)} = \sqrt{\int_a^b g_m^2(t)\, dt} \tag{6.7}$$

Orthonormal sets of functions satisfy $\|g_m\| = 1$ for every value of m.

Hermite polynomials are not orthogonal; however, they can be modified to become orthogonal as follows [66]:

$$h_n(t) = k_n e^{-t^2/4\tau^2} h_{e_n}(t)$$
$$= (-\tau)^n e^{t^2/4\tau^2} \tau^2 \frac{d^n}{dt^n} \left(e^{-t^2/2\tau^2} \right) \tag{6.8}$$

where $n = 0, 1, 2, \ldots$ and $-\infty < t < \infty$. The result is a set of orthogonal functions $h_n(t)$ which can be easily derived for all values of n. As examples we can write

$$h_0(t) = k_0 e^{-t^2/4\tau^2}$$

$$h_1(t) = k_1 \frac{t}{\tau} e^{-t^2/4\tau^2}$$

$$h_2(t) = k_2 \left[\left(\frac{t}{\tau} \right)^2 - 1 \right] e^{-t^2/4\tau^2}$$

$$h_3(t) = k_3 \left[\left(\frac{t}{\tau} \right)^3 - 3\frac{t}{\tau} \right] e^{-t^2/4\tau^2}$$

$$h_4(t) = k_4 \left[\left(\frac{t}{\tau}\right)^4 - 6\left(\frac{t}{\tau}\right)^2 + 3 \right] e^{-t^2/4\tau^2} \qquad (6.9)$$

$$h_5(t) = k_5 \left[\left(\frac{t}{\tau}\right)^5 - 10\left(\frac{t}{\tau}\right)^3 + 15\frac{t}{\tau} \right] e^{-t^2/4\tau^2}$$

$$h_6(t) = k_6 \left[\left(\frac{t}{\tau}\right)^6 - 15\left(\frac{t}{\tau}\right)^4 + 45\left(\frac{t}{\tau}\right)^2 - 15 \right] e^{-t^2/4\tau^2}$$

$$h_7(t) = k_7 \left[\left(\frac{t}{\tau}\right)^7 - 21\left(\frac{t}{\tau}\right)^5 + 105\left(\frac{t}{\tau}\right)^3 - 105\frac{t}{\tau} \right] e^{-t^2/4\tau^2}$$

$$h_8(t) = k_8 \left[\left(\frac{t}{\tau}\right)^8 - 28\left(\frac{t}{\tau}\right)^6 + 210\left(\frac{t}{\tau}\right)^4 - 420\left(\frac{t}{\tau}\right)^2 + 105 \right] e^{-t^2/4\tau^2}$$

The constants k_n, $n = 0, 1, \ldots$, determine the energy of the pulses. We call the functions derived in Equation (6.9) modified Hermite pulses (MHP), and it can be shown that their general formula is the following:

$$h_n(t) = k_n e^{-t^2/4\tau^2} n! \sum_{i=0}^{[n/2]} \left(-\frac{1}{2}\right)^i \frac{(t/\tau)^{n-2i}}{(n-2i)!\, i!} \qquad (6.10)$$

where $[n/2]$ denotes the integer part of $n/2$. If it is desired that each $h_n(t)$ pulse has an energy of E_n, then we can show that

$$k_n = \sqrt{\frac{E_n}{\tau n!\, \sqrt{2\pi}}} \qquad (6.11)$$

Figure 6.1(a) shows h_n for four values of n from 0 to 3. The scaling parameter is $\tau = 5 \times 10^{-11}$ s, and the pulses have unit energy. The time duration of all five pulses is about 0.6 ns. Using Equations (6.3)–(6.5) and (6.8) it can be shown that, disregarding the constants k_n, all MHPs satisfy the following differential equations:

$$\tau^2 \ddot{h}_n(t) + \left(n + \frac{1}{2} - \frac{1}{4}\frac{t^2}{\tau^2}\right) h_n(t) = 0 \qquad (6.12)$$

$$\tau \dot{h}_n(t) + \frac{t}{2\tau} h_n(t) = n h_{n-1}(t) \qquad (6.13)$$

$$h_{n+1}(t) = \frac{t}{2\tau} h_n(t) - \tau \dot{h}_n(t) \qquad (6.14)$$

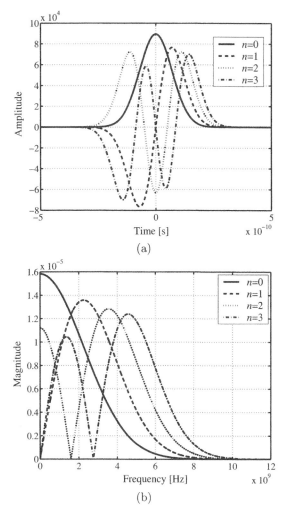

Fig. 6.1 Time and frequency responses of the normalized MHPs of orders $n = 0, 1, 2, 3$ normalized to unit energy.

Denoting the Fourier transform of $h_n(t)$ as $H_n(f)$, Equations (6.12)–(6.14) can be written as

$$\ddot{H}_n(f) + 16\pi^2\tau^2 \left(n + \frac{1}{2} - 4\pi^2 f^2\tau^2 \right) H_n(f) = 0 \tag{6.15}$$

$$j8\pi^2 f\tau^2 H_n(f) + j\dot{H}_n(f) = 4\pi n\tau H_{n-1}(f) \tag{6.16}$$

$$H_{n+1}(f) = j\frac{1}{4\pi\tau}\dot{H}_n(f) - j2\pi f\tau H_n(f) \tag{6.17}$$

respectively. Note that the derivatives are with respect to the frequency.

Example 6.1

Using Equation (6.17), find the frequency domain formulas for $h_1(t)$, $h_2(t)$, and $h_3(t)$.

Solution

For the value of $n = 0$, we will have $h_0(t) = k_0 e^{-t^2/4\tau^2}$ and

$$H_0(f) = k_0 \tau 2\sqrt{\pi}\, e^{-4\pi^2 f^2 \tau^2}$$

and from Equation (6.17) the transform of some higher degrees of MHPs can be obtained as follows (now we include the constants k_n):

$$H_1(f) = k_1 \tau (-j4\pi f\tau) 2\sqrt{\pi}\, e^{-4\pi^2 \tau^2 f^2}$$

$$H_2(f) = k_2 \tau \left(1 - 16\pi^2 f^2 \tau^2\right) 2\sqrt{\pi}\, e^{-4\pi^2 \tau^2 f^2} \qquad (6.18)$$

$$H_3(f) = k_3 \tau \left(-j12\pi f\tau + j64\pi^3 f^3 \tau^3\right) 2\sqrt{\pi}\, e^{-4\pi^2 \tau^2 f^2}$$

Figure 6.1(b) shows the frequency domain representations of the MHPs for $n = 0, 1, 2, 3$. According to Figures 6.1(a) and 6.1(b) all MHP pulses have the following properties:

- The pulse duration is almost the same for all values of n.

- The pulse bandwidth is almost the same for every value of n.

- The pulses are mutually orthogonal.

- The pulses have nonzero DC components, and, in fact, the low-frequency components of the pulses are relatively significant.

- The number of zero crossings is equal to n.

The autocorrelation functions of the MHPs for orders $n = 0, 1, 2, 3$ are shown in Figure 6.2. As can be seen, the width of the main peak in the autocorrelation function becomes narrower as the order of the pulses increases. This suggests that higher-order pulses will be more sensitive to pulse jitter.

6.1.3 Modulated and modified Hermite pulses

To gain more flexibility in the frequency domain, the time functions can be multiplied and modified by an arbitrary phase-shifted sinusoid. Hence, the multiplied and modified normalized Hermite pulses are defined as follows:

$$p_n(t) = \sqrt{2}\, h_n(t) \cos(2\pi f_c t + \phi_r) \qquad (6.19)$$

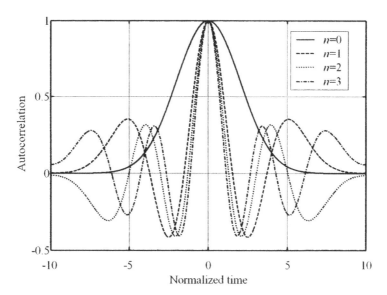

Fig. 6.2 Autocorrelation functions of modified normalized Hermite pulses of orders $n = 0, 1, 2, 3$. The width of the main peak in the autocorrelation function becomes narrower as the order of the pulse increases.

where $n = 0, 1, 2, \ldots$, $-\infty < t < \infty$, f_c is the shifting frequency, and ϕ_r is an arbitrary phase that can be zero without loss of generality. The function $p_n(t)$ for $n = 0, 1, 2, 3$ in the time and frequency domains is illustrated in Figures 6.3(a) and 6.3(b), respectively.

Comparing Figures 6.1 and 6.3 shows that $p_n(t)$ pulses have more oscillations than MHP pulses, but the pulse width has not been changed. The number of zero crossings in MHP is equal to n; however, in $p_n(t)$ it is a function of both n and f_c. We also observe that the fractional bandwidth of MHP is higher, and increasing f_c decreases the fractional bandwidth accordingly.

Equations (6.12) and (6.19) give the schematic diagram of a linear but time-variant system which creates pulses of different degrees. Figure 6.4 shows a simple schematic diagram for this system. The interesting feature of this structure is that, theoretically, by changing a simple gain of the circuit we can create different orders of the desired waveform.

According to Figures 6.3(a) and 6.3(b), all $p_n(t)$ functions have the following properties:

- The pulse duration is almost the same for all values of n.

- The pulse bandwidth is almost the same for every value of n.

- The fractional bandwidth can be easily controlled by f_c.

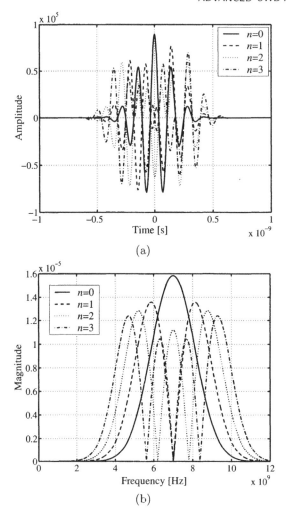

Fig. 6.3 Time and frequency representations of $p_n(t)$ for orders $n = 0, 1, 2, 3$. All pulses have zero low-frequency components. Compared with Figure 6.1(a), the number of zero crossings has been increased. It can also be seen that the fractional bandwidth of the signals has reduced from 200% to about 100% and can be further reduced by increasing f_c.

- The pulses are mutually orthogonal.

- The pulses have zero DC component.

6.2 ORTHOGONAL PROLATE SPHEROIDAL WAVE FUNCTIONS

In this section a multiple pulse generator which generates four different prolate spheroidal wave functions (PSWFs) based on a source signal for use in UWB

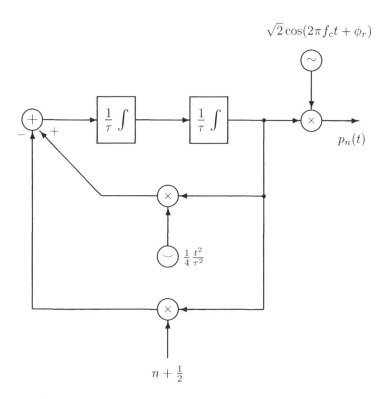

Fig. 6.4 The analog linear time-variant circuit producing different $p_n(t)$ functions.

communication systems is proposed. These sets of pulses can be applied to an M-ary communication system. This class of pulse shapes yields orthogonal pulses that have a constant pulse width and bandwidth regardless of the pulse order, which is in contrast to the majority of other orthogonal pulse classes. The results encourage the implementation of a cheap and easily reproducible UWB pulse generator.

6.2.1 Introduction

All conventional methods for enabling distinction between different pulse trains require that the multi-user base station send data serially, one pulse after another. Otherwise, interference between pulses will make it impossible for proper reception at the remote receivers. Therefore, encoding schemes become more complicated as the number of remote receivers increases; hence, the data rate per user decreases.

It is conceivable for a wireless impulse transmitter to communicate with multiple receivers using mutually orthogonal pulses, wherein a different pair of orthogonal pulses is assigned to each receiver to represent a digital signal [67]. Because all of the pulses are orthogonal, they can be transmitted simultaneously without interfering with each other. Moreover, since the transmitter and the receiver know in advance which pulses are for which receivers, each receiver will only receive signals including pulse shapes assigned to it. This would eliminate the need to modulate the pulse trains for identification purposes. However, with such a conceivable configuration, the transmitter would need to be provided with a separate pulse generator for each different pulse shape. The amount of hardware required for producing pulses would increase linearly with the number of receivers.

Considering the above-mentioned drawbacks, several objectives can be considered for a UWB transmitter:

- to provide a multipulse generator capable of producing two or more orthogonal pulses without much more hardware than required to produce a single pulse;

- to provide a wireless impulse transmitter with a simple configuration capable of high communication rates without the limitations caused by circuitry in PPM and the distance attenuation problem of PAM;

- to provide a wireless impulse system with a simple configuration that enables the number of users to be increased without reducing the communication rates.

In order to achieve the above-described objectives a multipulse generator for generating four different PWSF pulses, based on a source signal, is introduced. Conventionally, four complete sets of hardware are needed in order to produce four different pulses of different shapes. However, it will be shown how different pulses of different shapes and a desirable number of the pulse orders can be produced with a single-source signal, simply by changing a constant value. A wireless impulse transmitter is presented accordingly for transmitting pulse trains to a plurality of receivers which include a multipulse generator, a pulse selector, and a transmission unit. The pulse selector selects pulses from different pulse shape generators according to input data. The transmission unit transmits the pulses selected by the pulse selector.

6.2.2 Fundamentals of PSWFs

A function $h(t)$ is said to be time-limited if, for $T > 0$, $h(t)$ vanishes for all $|t| > T/2$. In an analogous manner, $H(f)$ is said to be band-limited with a bandwidth of Ω if it is zero outside of the band $(-\Omega, \Omega)$. Functions that are practically time- and band-limited can be very useful in communication engineering. One of these kinds of functions is referred to as a prolate spheroidal wave function. This function is the solution of [68]

$$\int_{-T/2}^{T/2} \psi_n(x) \frac{\sin \Omega(t-x)}{\pi(t-x)}\, dx = \lambda_n \psi_n(t) \tag{6.20}$$

or, alternatively, the solution of the differential equation

$$\frac{d}{dt}\left(1-t^2\right)\frac{d\psi_n(t)}{dt} + \left(\chi_n - c^2 t^2\right)\psi_n(t) = 0 \tag{6.21}$$

where $\psi_n(t)$ are PSWFs of order n and χ_n are the eigenvalues of $\psi_n(t)$. The constant c is

$$c = \frac{\Omega T}{2} \tag{6.22}$$

where Ω is the bandwidth and T is the pulse duration.

In Equation (6.20), λ_n is the concentration of energy in the interval $[-T/2, T/2]$ given by

$$\lambda_n = \frac{\int_{-T/2}^{T/2} |\psi_n(t)|^2 \, dt}{\int_{-\infty}^{\infty} |\psi_n(t)|^2 \, dt} \tag{6.23}$$

whose values range from 0 to 1. Moreover, it can be shown that PSWFs have the following properties [68]:

$$\int_{-\infty}^{\infty} \psi_m(t)\psi_n(t) \, dt = \begin{cases} 1, & m = n \\ 0, & m \neq n \end{cases} \tag{6.24}$$

$$\int_{-T/2}^{T/2} \psi_m(t)\psi_n(t) \, dt = \begin{cases} \lambda_n, & m = n \\ 0, & m \neq n \end{cases} \tag{6.25}$$

From Equations (6.24) and (6.25) a double orthogonality of $\psi_n(t)$ can be observed. This is a useful property for the UWB communication system because it guarantees unique demodulation at the receiver. This is the reason orthogonal pulse sets have been proposed for use in UWB applications. These orthogonal pulse sets consist of coded monocycles or coded baseband waveforms.

If we solve the differential equation (6.21) for the highest derivative, we get

$$\left(1 - t^2\right) \frac{d^2 \psi_n(t)}{dt^2} - 2t \frac{d\psi_n(t)}{dt} + \left(\chi_n - c^2 t^2\right) \psi_n(t) = 0 \tag{6.26}$$

and consequently

$$\ddot{\psi}_n(t) = \frac{1}{1 - t^2} \left[2t\dot{\psi}_n(t) - \left(\chi_n - c^2 t^2\right) \psi_n(t) \right] \tag{6.27}$$

As can be seen, different orders of the pulses can be simply obtained by changing the values of χ_n; hence, Equation (6.27) is the basis of the multipulse generator.

Different values of χ_n have been calculated and are shown in Table 6.1. (Details of the calculations can be found in [69].)

Here are some properties of PSWF wave shapes:

- The pulse duration is exactly the same for all values of n.

- The pulse bandwidth is almost the same for all values of n.

- The pulses are double-orthogonal.

Table 6.1 Eigenvalues of different pulse orders for $c = 2$.

Pulse order, n	χ_n
0	1.127 73
1	4.287 12
2	8.225 71
3	14.100 20
4	22.054 82
5	32.035 26
6	44.024 74
7	58.018 37
8	74.014 19
9	92.011 30

- The pulses have a nonzero DC component.

- Pulse duration and bandwidth can be controlled simultaneously.

If $\psi_n(t)$ is written in terms of the prolate angular function of the first kind,

$$\psi_n(t) = \psi_n(\Omega, T, t)$$
$$= \frac{[2\lambda_n(c)/T]^{1/2} S_{0n}^1(c, 2t/T)}{\left\{ \int_{-1}^{1} [S_{0n}^1(c, x)]^2 \, dx \right\}^{1/2}} \tag{6.28}$$

and

$$\left\{ \int_{-1}^{1} [S_{0n}^1(c, x)]^2 \, dx \right\}^{1/2} = \frac{2}{2n + 1} \tag{6.29}$$

where S_{0n}^1 is the prolate angular function of the first kind and λ_n is the fraction of the energy of $\psi_n(t)$ that lies in the interval $[-1, 1]$.

The prolate angular function of the first kind is given by [70]

$$S_{0n}^1(c, t) = \begin{cases} \displaystyle\sum_{k=0,2,\ldots}^{\infty} d_k(c) P_k(c, t) & \text{for } n \text{ even} \\[2em] \displaystyle\sum_{k=1,3,\ldots}^{\infty} d_k(c) P_k(c, t) & \text{for } n \text{ odd} \end{cases} \tag{6.30}$$

where $P_k(c, t)$ are the Legendre polynomials [71] and $d_k(c)$ are the series coefficients. Differentiation of $S_{0n}^1(c, t)$ is given by

$$\dot{S}_{0n}^1(c, t) = \sum_{k} d_k(c) \dot{P}_k(c, t) \tag{6.31}$$

If the associated Legendre functions are given in terms of unassociated Legendre functions, then [72]

$$P_k^m(t) = (-1)^m \left(1 - t^2\right)^{1/2} \frac{d^m}{dt^m} P_k(t) \tag{6.32}$$

and for the first derivative

$$\frac{dP_k(t)}{dt} = \frac{-P_k^1(t)}{\sqrt{1-t^2}} \tag{6.33}$$

using Equations (6.31)–(6.33) yields

$$\dot{S}_{0n}^1(c,t) = \frac{-1}{\sqrt{1-t^2}} \sum_k d_k(c) P_k^1(t) \tag{6.34}$$

and

$$P_k^1(t) = \frac{-1}{2^k k!} \left(1 - t^2\right)^{1/2} \frac{d^{k+1}}{dt^{k+1}} \left(t^2 - 1\right)^k \tag{6.35}$$

The second differentiation of $S_{0n}^1(c,t)$ is given by

$$\ddot{S}_{0n}^1(c,t) = \sum_k d_k(c) \ddot{P}_k(c,t) \tag{6.36}$$

Using Equation (6.32) yields

$$\frac{d^2 P_k(t)}{dt^2} = \frac{P_k^2(t)}{1 - t^2} \tag{6.37}$$

Using Equations (6.36) and (6.37) the following results:

$$\ddot{S}_{0n}^1(c,t) = \frac{1}{1-t^2} \sum_k d_k(c) P_k^2(t) \tag{6.38}$$

and

$$P_k^2(t) = \frac{1}{2^k k!} \left(1 - t^2\right) \frac{d^{k+2}}{dt^{k+2}} \left(t^2 - 1\right)^k \tag{6.39}$$

6.2.3 PSWF pulse generator

A UWB communication system employing a PSWF pulse generator, a wireless UWB impulse transmitter, and two remote receivers can be constructed as shown in Figure 6.5. The transmitter is a multi-user base station and the receivers 1 and 2 are mobile terminals.

The transmitter includes a pulse source, four pulse generators, a pulse train selector/combiner, and a transmission unit. Pulse generators are individually connected to the pulse selector/combiner. The four pulse generators generate PSWF pulses of orders 2 to 5, respectively, and are in synchronization based on the inputs from pulse source 2. The pulse selector/combiner selects pulses from pulse generators based on

Data inputs for users 1 and 2

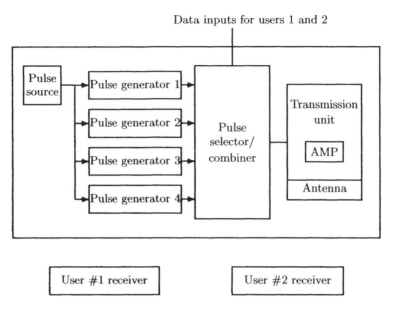

Fig. 6.5 Schematic diagram of a UWB communication system employing a PSWF pulse generator (AMP, amplifier).

a two-to-one correspondence between the different-order pulses and receivers 1 and 2; that is, the pulse selector/combiner selects pulses from pulse generators 1 and 2 based on input data to be sent to the receiver 1 and selects pulses from the pulse generators 3 and 4 based on input data to be sent to receiver 2. This is because the four pulse shapes generated by pulse generators are represented by symbol numbers 1 to 4 and each of receivers 1 and 2 is assigned two of the four pulse shapes (symbol numbers) to represent a binary channel, that is, 0 or 1. In this example, receiver 1 is assigned symbol numbers 1 and 2 and receiver 2 is assigned symbol numbers 3 and 4. This correspondence relationship is summarized in Table 6.2.

Table 6.2 The relationship between receivers and assigned symbols.

Receiver	Binary	Symbol
1	0	1
1	1	2
2	0	3
2	1	4

The pulse selector/combiner also combines the pulses for receivers 1 and 2 and sends them to the transmission unit. The transmission unit amplifies the pulses using the amplifier and transmits them over the antenna. The transmission unit transmits pulses corresponding to different receivers simultaneously.

The configuration of pulse generators is illustrated in Figure 6.6. It should be noted that pulse generators are designed based on Equation (6.27). The results of PSWF generation are illustrated in **Figure 6.7** for orders 4 and 5.

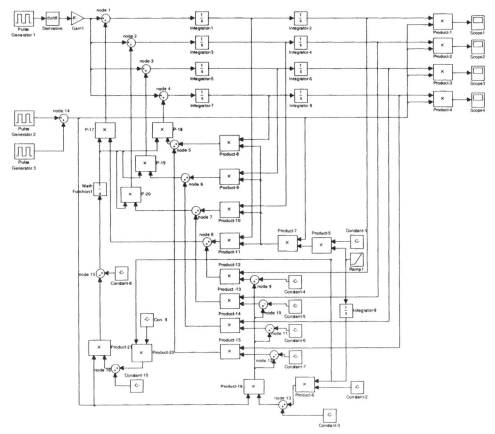

Fig. 6.6 Schematic diagram of four different PSWF pulse generators.

Each of the first-order to fourth-order pulse generators has the same configuration for generating PSWF orthogonal pulses. In this way, four different-order pulses can be generated with only slightly more hardware than required to produce one PSWF orthogonal pulse.

Each of receivers 1 and 2 has substantially the same configuration in order to demodulate incoming signals, with the exception of the two pulse shapes assigned to each. Here, an explanation will be provided for receiver 1 as a representative example. As shown in Figure 6.8, receiver 1 includes a reception unit W, a timing control circuit, a basic pulse supplier, a correlator, an orthogonal-to-digital data selector, and an output unit. The reception unit W includes an antenna, a filter, and an amplifier.

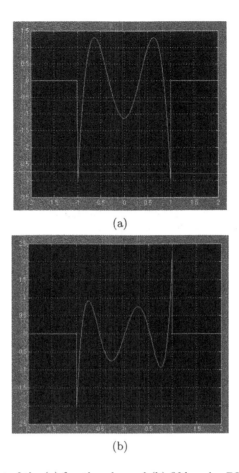

(a)

(b)

Fig. 6.7 Output of the (a) fourth-order and (b) fifth-order PSWF generators.

After receiving a pulse by the antenna the received signal is filtered and amplified. The basic pulse supplier supplies two different-order pulse waveforms and, in the present embodiment, the second- and third-order pulse waveforms.

The correlator measures the similarity between each incoming pulse and the plurality of orthogonal pulse shapes from the basic pulse supplier to identify the corresponding symbol. Because orthogonal pulses are used the cross-correlation between different-order pulses is zero. Therefore, the correlator can correctly distinguish among different-order pulses. The correlator performs the correlation process at a timing modified by the timing control circuit to allow for differences in time of flight between the transmitter and receiver 1; for example, this may be when the transmitter and receiver 1, or both, are moved. It also allows inclusion of PPM and PN code timing changes. In the example shown in Figure 6.8, the correlator determines that four successive incoming pulses correspond to the symbol numbers 1, 2, 2, 1.

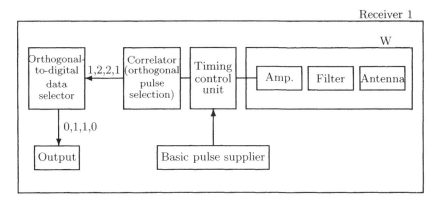

Fig. 6.8 Schematic diagram of the PSWF UWB receiver.

The orthogonal-to-digital data selector converts symbol numbers from the correlator into corresponding binary data. In Figure 6.8, the orthogonal-to-digital data selector converts the symbol numbers 1, 2, 2, 1 from the correlator into corresponding binary data of 0, 1, 1, 0. Receivers 1 and 2 have the ability to demodulate all the different orthogonal pulse shapes assigned to the system. This enables implementation of an M-ary modulation scheme when only one or other of receivers 1 and 2 are being used, to enhance the transmission rate of the system. For example, when only receiver 1 is used the transmitter assigns all of the different-order pulse waveforms to receiver 1 and assigns each pulse shape correspondence with multi-bit symbols as indicated in Table 6.3 to create a 4-ary modulation scheme.

Table 6.3 Creation of a 4-ary modulation scheme.

Receiver	Binary	Symbol
1	00	1
1	01	2
1	10	3
1	11	4

Although we have described only four pulse generators, six or more different-order pulse generators can be provided as well. In such a case, each pair of pulse generators could be assigned to a different receiver in order to increase the number of receivers that can be used within the system. Alternatively or in addition, each pair of generators could be assigned a different multi-bit symbol to increase the transmission rate of the system as described above. The transmission rate of the system could be enhanced by further use of conventional modulation schemes, such as PPM or PAM.

6.3 WAVELET PACKETS IN UWB PSM

This section explains wavelet packets for use in UWB communications. The pulse shapes that are generated are nearly orthogonal and have almost the same time duration. After normalization, an M-ary signaling set can be constructed using these properties, allowing a higher data rate.

Wavelet analysis is a relatively new field of mathematical research which has generated great interest in theoretical and applied mathematics over the past decade [73, 74]. Let $L^2(a, b)$ denote the collection of measurable functions $f(x)$, defined on the interval (a, b), that satisfy

$$\int_a^b |f(x)|^2 \, dx < \infty \tag{6.40}$$

Assume that $\phi \in L^2(\mathbb{R})$ is the scaling function corresponding to the wavelet $\psi \in L^2(\mathbb{R})$. If we define $\psi_{j,k}$ by

$$\psi_{j,k} = 2^{j/2} \psi(2^j x - k) \tag{6.41}$$

where $j, k \in \mathbb{Z}$, then ψ is called a semi-orthogonal wavelet if $\{\psi_{j,k}\}$ satisfies

$$\langle \psi_{j,k}, \psi_{l,m} \rangle = 0, \quad j \neq l;\ j, k, l, m \in \mathbb{Z} \tag{6.42}$$

Moreover, if $\{\psi_{j,k}\}$ satisfies

$$\langle \psi_{j,k}, \psi_{l,m} \rangle = \delta_{j,l} \delta_{k,m}, \quad j, k, l, m \in \mathbb{Z} \tag{6.43}$$

then ψ is called an orthogonal wavelet. Finally, ϕ is described by a unique sequence $\{p_k\}$ as follows:

$$\phi(x) = \sum_{k=-\infty}^{\infty} p_k \phi(2x - k) \tag{6.44}$$

where $\{p_k\}$ is called the two-scale sequence of ϕ. Then, the two-scale relationship of the corresponding wavelet ψ is given by

$$\psi(x) = \sum_{k=-\infty}^{\infty} q_k \phi(2x - k) \tag{6.45}$$

where

$$q_n = (-1)^n \bar{p}_{-n+1} \tag{6.46}$$

and '$\bar{\ }$' denotes the reversal or transpose operator. Let $\{p_k\}$ be the two-scale sequence of an orthogonal scaling function ϕ, and $\{q_k\}$, as defined in Equation (6.46), be the two-scale sequence which completely characterizes its corresponding orthogonal wavelet ψ. If we note that

$$\begin{cases} \mu_0(x) = \phi(x) \\ \mu_1(x) = \psi(x) \end{cases} \tag{6.47}$$

we can define the family of functions μ_n, $n = 2l$ or $2l + 1$, $l \in \mathbb{Z}_+$, defined by

$$\begin{cases} \mu_{2l}(x) = \sum_k p_k \mu_l(2x - k) \\ \mu_{2l+1}(x) = \sum_k q_k \mu_l(2x - k) \end{cases} \tag{6.48}$$

These functions are called wavelet packets relative to the orthogonal scaling function $\mu_0 = \phi$ and satisfy the following orthogonality properties:

$$\langle \mu_n(\cdot - j), \mu_n(\cdot - k) \rangle = \delta_{j,k}, \quad j, k, n \in \mathbb{Z} \tag{6.49}$$
$$\langle \mu_{2l}(\cdot - j), \mu_{2l+1}(\cdot - k) \rangle = 0, \quad j, k \in \mathbb{Z}, \ l \in \mathbb{Z}_+ \tag{6.50}$$

Furthermore, these wavelet packets must be finite energy functions with Lebesgue measure, i.e. the Lebesgue integral as defined in Equation (6.40) exists.

Using an orthogonal or semi-orthogonal wavelet, chosen in a wavelet family such as Haar, biorthogonal, Coiflet, Daubechies, or Symlet, we can easily generate the corresponding wavelet packets, as defined in Equation (6.48). For each set thus constructed, we can notice that the wavelets have almost the same time duration. However, except the set using the Haar wavelet, the orthogonality of these sets is not exact, but for practical applications it might be adequate. Concerning the orthogonality of the set using the Haar wavelet, this advantage has to be minimized by the fact that the Haar wavelet is irregular and consequently difficult to generate in practice.

Regarding the frequency domain of these wavelet packets, the pulse bandwidth is almost the same for each wavelet. However, each wavelet has its own center frequency, and the fractional bandwidth takes its values in the interval [20%, 200%]. This property is satisfied for each set and that puts in evidence the fact that it might be difficult to find an antenna that can transmit all these pulses because of the large frequency band that all these sets require.

It is interesting to observe that the set constructed using the Daubechies order 5 wavelet has almost the same frequency components compared with that using the Symlet order 5 wavelet. By using wavelets in both sets we can obtain a new set that requires less bandwidth. In Figures 6.9 and 6.10 the time and frequency domains of the normalized sets using wavelets in both Daubechies and Symlet wavelet packets are shown. Note that this set does not respect the ruling by the FCC which allows UWB signals with a peak PSD value of -41.3 dBm/MHz between 3.1 GHz and 10.6 GHz. However, it is possible to define a set that respects this ruling by considering less symbols, using another wavelet family, or dilating the wavelets. This implies a tradeoff between the pulse width (related to the data rate or system capacity), the adjacent channel interference in the frequency domain, and the desired number of orthogonal pulses.

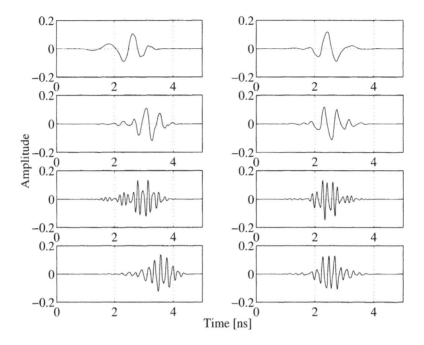

Fig. 6.9 Time domain of normalized sets using wavelets in both Daubechies and Symlet wavelet packets. The time duration of each wavelet is almost the same. From top to bottom are represented $\mu_1, \mu_3, \mu_5, \mu_7$ according to Daubechies and Symlet packets, on the left and right, respectively.

6.3.1 PSM system model

Now let us consider a PSM technique for our UWB system. Using the Daubechies order 5 and Symlet order 5 sets described in Figure 6.9, we can easily construct an M-ary signaling system, where M is the number of symbols. The baseband representation of a transmitted sequence is given by

$$s(t) = \sum_{n=-\infty}^{\infty} w_{i,n}(t - nT_f) \qquad (6.51)$$

In the above equation, $w_{i,n}$ is the nth transmitted symbol where $i \in [1, M]$ denotes the ith symbol of the M-ary set. The frame period is represented by T_f. In order to combat the ISI due to the multipath fading environment, a guard time T is inserted between each transmitted symbol. If we denote by T_s the time duration of a symbol, then the frame period is given by $T_f = T_s + T$. The bit rate R_b is then defined by

$$R_b = \frac{\log_2(M)}{T_f} \qquad (6.52)$$

As an example, M can be 8, assuming three bits per symbol, and T_s can be a couple of nanoseconds.

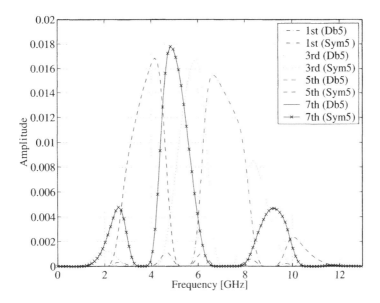

Fig. 6.10 Frequency domain of the normalized set using wavelets in both Daubechies and Symlet wavelet packets.

6.3.2   Receiver structure

In order to investigate the performance of the system, we assume that the transmitted sequence as in Equation (6.51) propagates through a multipath channel with impulse response

$$h(t) = \sum_{l=1}^{L} \alpha_l \delta(t - \tau_l) \tag{6.53}$$

where α_l and τ_l are the amplitude and the delay of the lth path, respectively, and L is the total number of paths characterizing the channel.

Let us now define $g_{i,n}(t)$ as

$$g_{i,n}(t) = w_{i,n}(t) * h(t) \tag{6.54}$$

where $*$ denotes the convolution. The received signal at the output of the receiver antenna is then

$$r(t) = \sum_{n=-\infty}^{\infty} g_{i,n}(t - nT_f - \tau_l) + n(t) \tag{6.55}$$

where $n(t)$ represents the receiver zero-mean AWGN with two-sided PSD of $N_0/2$.

As there exist many resolvable paths in the UWB multipath channel as in Equation (6.53), we then consider a rake receiver structure to correlate the received signal with the M template waveforms w_j, $j \in [1, M]$. Basically, the rake receiver tracks the L_p best paths out of the L available paths, and information is then detected through MRC. We can assume that the receiver has perfect knowledge of the propagation channel, and it is perfectly synchronized to the received signal [75].

6.4 SUMMARY

This chapter discussed the various aspects and features of advanced pulse generation in UWB communication systems. We started with a class of mathematical functions called Hermite polynomials. They were modified and made orthogonal. The fundamental equations representing the time and frequency domain characteristics of these functions were derived. We observed that a linear time-variant circuit can generate the pulses of different orders simultaneously.

As a second example of orthogonal sets of pulse shapes, we defined the PSWFs. A generator for producing four different orders of these functions in a UWB communication system was designed. Moreover, the pulses were applied to an M-ary transmission system and details for employment of the pulse generators were explained.

We explained nearly orthogonal wavelet packets for use in UWB communications. The pulse shapes had almost the same time duration and, after normalization, an M-ary signaling set was constructed using these properties allowing higher data rates.

Problems

Problem 1. Using the general formula in Equation (6.10) investigate the validity of the MHP functions written in Equation (6.9).

Problem 2. By using the condition for orthogonality expressed in Equation (6.6) show that $h_0(t)$, $h_1(t)$, $h_2(t)$, and $h_4(t)$ are mutually orthogonal. Are they orthonormal as well?

Problem 3. Verify the validity of the formulas for $H_1(f)$, $H_2(f)$, and $H_3(f)$ in Equation (6.18). Find a proper value for τ that gives a 3 dB bandwidth of 3 GHz for $H_2(f)$.

Problem 4. Assume that MHP pulse shapes have been applied to a differentiating process. Is their orthogonality maintained in spite of the differentiation?

Problem 5. Verify the eigenvalues χ_n given in Table 6.1. Use the details of the calculations in [69].

Problem 6. Show that

$$\sum_{n=0}^{\infty} \psi_n(0)\psi_n(t) = \text{sinc}(t) \tag{6.56}$$

Problem 7. Show that

$$\sum_{n=0}^{\infty} \psi_n(k)\psi_n(m) = \delta_{km} \tag{6.57}$$

Problem 8. Show that

$$\sum_{k=-\infty}^{\infty} \psi_n(t-k) = \int_{-\infty}^{\infty} \psi_n(x)\,dx = \hat{\psi}_n(0) \tag{6.58}$$

7

UWB antennas and arrays

Ultra short-pulse technologies are today generally considered practical only for low-power and short-range applications due to the limitations of pulse-forming electronics; however, this state of affairs is quickly changing. Building UWB communication systems requires a theoretical basis for computation and estimation of antenna design parameters and performance prediction. This is particularly important for the design of antenna systems that determine the performance of precision range and direction measurements. In fact, antenna design is an important UWB RF challenge. While wide bandwidth types of antennas are well understood in other applications, such as radar, UWB communication systems require an antenna with a flat group delay and a small size, so that the high- and low-frequency signal components arrive at the receiver simultaneously and the antenna can fit into consumer electronics products such as digital cameras and camcorders. It is expected that an appropriate antenna configuration should be part of a UWB chipset's reference design.

This chapter is about UWB antennas and arrays. These kinds of antennas are specifically designed to transmit and receive very short-time duration pulses of electromagnetic energy. An impulse antenna requires consideration of some concepts that may be new to many readers. The antenna's components act as distributed reactances and change the shape of the incoming waveforms that excite the antenna. Beamforming techniques for nonsinusoidal UWB signals are also considered in this chapter. In addition, radar UWB array systems and their properties are briefly discussed. It is believed that UWB antenna design remains one of the major issues in the progress of UWB technology.

Ultra Wideband Signals and Systems in Communication Engineering   Second Edition
M. Ghavami, L. B. Michael and R. Kohno   © 2007 John Wiley & Sons, Ltd

7.1 ANTENNA FUNDAMENTALS

An antenna is a device that transmits and/or receives electromagnetic waves. Electromagnetic waves are often referred to as radio waves. Most conventional antennas
are resonant devices that operate efficiently over a relatively narrow frequency band.
An antenna must be tuned to the same frequency band as the radio system to which
it is connected operates in, otherwise reception and/or transmission will be impaired.
In this section we discuss some fundamentals of antenna systems, such as Maxwell's
equations, far-field and near-field regions, antenna patterns, and antenna gain.

7.1.1 Maxwell's equations for free space

Maxwell's equations represent one of the most elegant and concise ways of stating the
fundamentals of electricity and magnetism. From them we can develop most of the
working relationships in the field. Because of their conciseness they embody a high
level of mathematical sophistication and are therefore not generally introduced in an
introductory treatment of the subject, except perhaps as summary relationships.

Maxwell's equations tell us that a time-varying current can radiate as a free space
electromagnetic field. The purpose of any antenna is to act as a launching means
between guided and free space electromagnetic waves or vice versa. The 'guiding'
might be in the form of a wire, a coaxial cable, a stripline, or an actual waveguide.

All four Maxwell equations are available in two forms of integration and differentiation [76].

7.1.1.1 Gauss's law for electricity The electric flux out of any closed surface is
proportional to the total charge enclosed within the surface

$$\text{Integral form:} \qquad \oint_A \mathbf{E} \cdot d\mathbf{A} = \frac{Q}{\epsilon_0} \tag{7.1}$$

where $\mathbf{E}$ is the electric field (in units of V/m), $d\mathbf{A}$ is the area of a differential square
on the surface A with an outward-facing surface normal defining its direction, Q is the
charge enclosed by the surface, and ϵ_0 (approximately 8.854 pF/m) is the permittivity
of free space. Note that the integral form only works if the integral is over a closed
surface.

The differential form (Equation (7.1)) can be written as

$$\text{Differential form:} \qquad \nabla \cdot \mathbf{E} = \frac{\rho}{\epsilon_0} \tag{7.2}$$

where $\nabla \cdot \mathbf{E}$ is the divergence of $\mathbf{E}$ and ρ is the charge density.

The integral form of Gauss's law finds application in calculating electric fields
around charged objects. By applying Gauss's law to the electric field of a point charge
we can show that it is consistent with Coulomb's law. While the area integral of
the electric field gives a measure of the net charge enclosed, the divergence of the
electric field gives a measure of the density of sources. It also has implications for the
conservation of charge.

7.1.1.2 Gauss's law for magnetism The net magnetic flux out of any closed surface is zero

$$\text{Integral form:} \qquad \oint_A \mathbf{B} \cdot d\mathbf{A} = 0 \qquad (7.3)$$

where $\mathbf{B}$ is the net magnetic flux (in units of tesla, T). Similar to Equation (7.2) we can write

$$\text{Differential form:} \qquad \nabla \cdot \mathbf{B} = 0 \qquad (7.4)$$

These equations are related to the magnetic field's structure because it states that, given any volume element, the net magnitude of vector components that point outward from the surface must be equal to the net magnitude of vector components that point inward. Structurally, this means that magnetic field lines must be closed loops.

Another way of putting it is that field lines cannot originate from somewhere; attempting to follow the lines backward to their source or forward to their terminus ultimately leads back to the starting position. This implies that there are no *magnetic monopoles*.

7.1.1.3 Faraday's law of induction The line integral of the electric field around a closed loop is equal to the negative of the rate of change of the magnetic flux through the area enclosed by the loop

$$\text{Integral form:} \qquad \oint_L \mathbf{E} \cdot d\mathbf{l} = -\frac{\partial \phi}{\partial t} \qquad (7.5)$$

where $d\mathbf{l}$ is the differential length vector on the closed loop L and ϕ is the magnetic flux through the area A enclosed by the loop. The relation between ϕ and $\mathbf{B}$ can be written as

$$\phi = \int_A \mathbf{B} \cdot d\mathbf{A} \qquad (7.6)$$

This equation only works if the surface A is not closed, because the net magnetic flux through a closed surface will always be zero. The line integral of Equation (7.5) is equal to the generated voltage or electromotive force (emf) in the loop, so Faraday's law relates electric and magnetic fields and is the basis for electric generators. It also forms the basis for inductors and transformers.

Note that the negative sign is necessary to maintain conservation of energy. It is so important that it even has its own name, Lenz's law.

The differential form of Faraday's law of induction can be expressed as

$$\text{Differential form:} \qquad \nabla \times \mathbf{E} = -\frac{\partial \mathbf{B}}{\partial t} \qquad (7.7)$$

where $\nabla \times \mathbf{E}$ is the curl of $\mathbf{E}$.

7.1.1.4 Ampère's law The line integral of the magnetic field around a closed loop is comprised of two components. The first one is proportional to the electric current flowing through the loop and the second one is proportional to the rate of time variations of the integral of the electric flux out of the open surface A. This law is useful for calculation of the magnetic field for simple geometries

$$\text{Integral form:} \qquad \oint_L \mathbf{B} \cdot dl = \mu_0 I + \mu_0 \epsilon_0 \frac{\partial}{\partial t} \int_A \mathbf{E} \cdot d\mathbf{A} \qquad (7.8)$$

where I is total electric current and μ_0 is the permeability of the free space defined to be exactly $4\pi \times 10^{-7}$ W/(A m). Usually the second term on the right-hand side is negligible and ignored, so hence the integral form is known as *Ampère's law*.

The differential form of this law can be derived as follows

$$\text{Differential form:} \qquad \nabla \times \mathbf{B} = \mu_0 \mathbf{j} + \mu_0 \epsilon_0 \frac{\partial \mathbf{E}}{\partial t} \qquad (7.9)$$

where $\mathbf{j}$ is the current density.

7.1.2 Wavelength

Generally, the antenna size is referred relative to the signal wavelength if we are dealing with a narrowband system. For example, a half-wave dipole is approximately half a wavelength long. Wavelength is the distance a radio wave travels during one cycle. The formula for wavelength is

$$\lambda = \frac{c}{f} \qquad (7.10)$$

where λ is the wavelength, expressed in units of length, c is the speed of light in the medium, and f is the frequency.

For UWB signals we do not have a single operating frequency, and hence wavelengths of these sorts of signals are usually expressed as the lower and upper available wavelengths in the system.

7.1.3 Antenna duality

Most antennas work equally well in both directions, being able to transmit or receive signals. The mathematical fundamentals of antennas work equally well either way. The only limits for duality of antenna are power and overload restrictions.

7.1.4 Impedance matching

For efficient transfer of energy, impedance of the radio, antenna, and transmission line connecting the radio to the antenna must be the same. For instance, radios typically are designed for 50 Ω impedance and the coaxial cables (transmission lines) used with them also have a 50 Ω impedance. Efficient antenna configurations often have an impedance other than 50 Ω, and some sort of impedance-matching circuit is therefore required to transform the antenna impedance to 50 Ω.

7.1.5 Voltage standing wave ratio and reflected power

The voltage standing wave ratio (VSWR) is a well-known indication of how good the impedance match is. The VSWR is often abbreviated as SWR. A high VSWR is an indication that the signal is reflected prior to being radiated by the antenna. VSWR and reflected power are different ways of measuring and expressing the same thing.

A VSWR of 2:1 or less is considered good. Most commercial antennas, however, are specified to be 1.5:1 or less over some bandwidth. Based on a 100 W radio, a 1.5:1 VSWR equates to a forward power of 96 W and a reflected power of 4 W, or the reflected power is 4.2% of the forward power.

7.1.6 Antenna bandwidth

Bandwidth can be defined in terms of radiation patterns or VSWR/reflected power. Bandwidth is often expressed in terms of percent or fractional bandwidth (FB), because the percent bandwidth is constant relative to frequency. In this case, if the bandwidth is expressed in absolute units of frequency (e.g., gigahertz) the fractional bandwidth is then different depending upon whether the frequencies in question are near 3, 4, or 8 GHz.

By definition, the fractional bandwidth of a signal is the ratio of the bandwidth to the center frequency as follows:

$$\text{FB} = \frac{f_h - f_l}{(f_h + f_l)/2} \times 100\% \tag{7.11}$$

where f_h and f_l are the effective highest and lowest frequency components of the signal, respectively. Wideband arrays are designed with fractional bandwidth of up to 20% and UWB arrays are proposed with fractional bandwidth of 20 to 200%.

7.1.7 Directivity and gain

Directivity is the ability of an antenna to focus energy in a particular direction when transmitting or to receive energy in a better way from a particular direction when receiving. The relationship between gain and directivity is based on efficiency,

$$\text{Gain} = \frac{\text{Efficiency}}{\text{Directivity}} \tag{7.12}$$

We can see the phenomena of increased directivity by comparing a light bulb with a spotlight. A 100 W spotlight will provide more light in a particular direction than a 100 W light bulb and less light in other directions. We could say the spotlight has more 'directivity' than the light bulb. The spotlight is comparable with an antenna with increased directivity. An antenna with increased directivity, that is hopefully implemented efficiently, is low loss, and, therefore, exhibits both increased directivity and gain.

Gain is given in reference to a standard antenna. The two most common reference antennas are the *isotropic antenna* and the *resonant half-wave dipole antenna*.

The isotropic antenna radiates equally well in all directions. Real perfect isotropic antennas do not exist, but they provide useful and simple theoretical antenna patterns with which to compare real antennas. An antenna gain of 2 (3 dB) compared with an isotropic antenna would be written as 3 dBi. The resonant half-wave dipole can be a useful standard for comparison with other antennas at one frequency or over a very narrow band of frequencies. To compare the dipole with an antenna over a range of frequencies requires an adjustable dipole or a number of dipoles of different lengths. An antenna gain of 1 (0 dB) compared with a dipole antenna would be written as 0 dBd.

7.1.8   Antenna field regions

Most antennas have two operating regions which are called the *near-field* and the *far-field* regions. These are sometimes called the Fraunhofer and Fresnel regions. In the antenna near-field, behavior is highly complex and most energy drops off with the cube of distance. In the antenna far-field, properties are more orderly and most energy falls off with the square of the distance. The crossover between near-field and far-field takes place at $2L^2/\lambda$ or around a wavelength for a normal antenna, where L is the antenna width. Figure 7.1 shows how the field strength drops with distance on most typical antennas.

7.1.9   Antenna directional pattern

The directional (radiation) pattern of an antenna or an array of antenna elements describes the relative strength of the radiated field in various directions from the antenna or array, at a fixed or constant distance. The radiation pattern is a 'reception pattern' as well, since it also describes the receiving properties of the antenna.

The radiation pattern is three-dimensional, but it is difficult to display in a meaningful manner; it is also time-consuming to measure a three-dimensional radiation pattern. Often, radiation patterns are a slice of the three-dimensional pattern, which is of course a two-dimensional radiation pattern that can be displayed easily on a screen or piece of paper. These pattern measurements are presented in either a rectangular or a polar format.

An antenna pattern defined by uniform radiation in all directions and produced by an isotropic radiator (or a point source: a nonphysical antenna which is the only nondirectional antenna) is called an *isotropic pattern*. A pattern which is uniform in a given plane is called an *omni-directional pattern*.

Generally, it is desired that an antenna sends or gathers in energy in a particular direction, thereby creating a special antenna radiation pattern. Patterns can be created by a single antenna element or by the arrangement and size of a set of antenna elements. For instance, a TV satellite antenna must have a very narrow beamwidth, because the desired satellite has a weak signal in any other direction than that directly toward the satellite. However, the antenna of a GPS navigation receiver

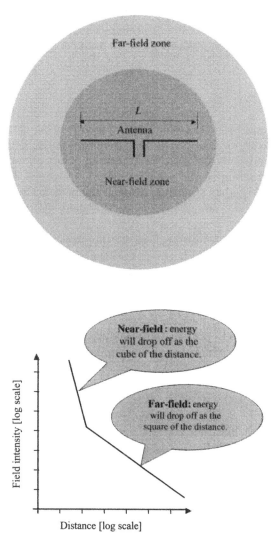

Fig. 7.1 Typical antennas have near-field and far-field regions. The behavior of the two regions is radically different. Near-field mathematics is quite complex, whereas far-field mathematics is more orderly.

usually has to follow six moving satellites at once. Hence, the antenna should have a half-hemispherical pattern that can receive equally well from any point in the sky.

Terrestrial TV transmitters create a 'bagel' pattern since transmitting energy up toward the sky or into the earth is wasteful. Multitower AM broadcast antennas purposely design nulls toward neighboring stations to lower interference.

Directional patterns of wideband or UWB antennas and arrays are always functions of frequency. Sophisticated antenna design techniques and signal-processing algorithms have to be proposed in order to solve the problem of the frequency dependence of antennas and arrays.

7.1.10 Beamwidth

Depending on the radio system in which an antenna is being employed, there can be many definitions of beamwidth. A common definition is the half-power beamwidth. The peak radiation intensity is found, and then the points on either side of the peak representing half the power of the peak intensity are located. The angular distance between the half-power points travelling through the peak is the beamwidth. Half the power is 3 dB, so the half-power beamwidth is sometimes referred to as the 3 dB beamwidth.

Inter-null beamwidth can also be used as the measure of the directivity of the antenna. By definition, inter-null beamwidth is the difference between the two angles corresponding to the first nulls on either side of the peak radiation.

Figure 7.2 shows the directional pattern parameters of an antenna (array). As indicated in this figure, the *main lobe*, which is also called the *major lobe* or *main beam*, is the radiation lobe in the direction of maximum radiation. The *side lobe* is a radiation lobe in any direction other than the direction(s) of the intended radiation. The *back lobe* is defined as the radiation lobe opposite the main lobe.

7.2 ANTENNA RADIATION FOR UWB SIGNALS

The radiation for short-duration UWB signals from an antenna is significantly different compared with the radiation produced by long-duration narrowband signals. These differences are mostly due to the following parameters [77]:

- the characteristics of the radiation of the signal from the antenna aperture;

- the time domain attributes of the radiation field;

- the amplitude of the radiation field;

- the spatial attributes of the radiation field;

- the properties of side radiation.

The development of new UWB sources and antennas has shown significant progress in recent years. In particular, research in impulse-radiating antennas has improved the performance of a number of other wideband systems. The development of a theoretical description of impulse-radiating antennas has been considered for some years, and theoretical models are now available for far-field and near-field radiations of various antenna types (e.g., reflector, horns, and antenna arrays).

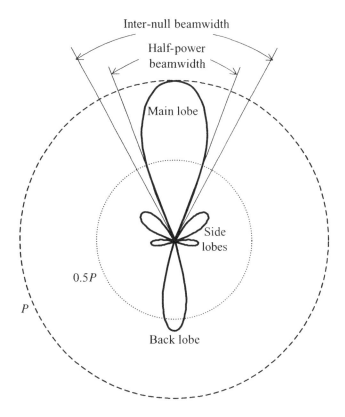

Fig. 7.2 Antenna directional pattern parameters. It is assumed that the power at the desired direction is P. Hence, the half-power circle is identified by $0.5P$ and determines the half-power beamwidth.

Due to the strong impact of the waveform on the coupling of transient fields into a system, the parameters of the waveform, such as rise time or pulse duration, are often more important than the pulse amplitude for susceptibility investigations. Based on the limited amplitude range of high-power pulse sources, field magnitudes are often scaled by changing the distance between the antenna and target or by changing the direction angle. The radiated impulse shape depends on the direction angle (e.g., rise time decreases for increasing direction angles).

We start with a short review of an analytical model for the radiation of aperture antennas. Following this a review of the frequency domain far-field definition is performed. Based on these theoretical reviews a time domain far-field definition will be developed with respect to waveform effects. Comparison of the developed definition and the classical frequency domain definitions shows that for fast transient radiation the well-known frequency domain estimation models fail because the dimensions of antenna systems and wave propagation lead to a special kind of dispersion. The importance of this near-field dispersion for practical applications has been demonstrated by both theoretical considerations and measurements with a half-impulse radiation antenna and an 8×8 antenna array [78].

The radiation behavior of most impulse-radiating antennas can be represented by the electrical field ($\mathbf{E}_a$) inside the aperture (A_a) using the equivalence principle [79]. In the frequency domain the electrical field strength at an arbitrary observation point r_p can be calculated by the integral

$$\mathbf{E}_s(\mathbf{r}_p, s) = \frac{1}{2\pi} \int_{A_a} \frac{sR + c}{cR^2} [(\mathbf{e}_z \times \mathbf{E}_a(\mathbf{r}_q, s)) \times \mathbf{e}_\mathbf{R}] e^{-sR/c} \, dA_a \qquad (7.13)$$

where $\mathbf{e}_z$ and $\mathbf{e}_\mathbf{R}$ are the unit vectors of the z-axis and $\mathbf{R}$, respectively. The complex frequency is denoted by $s = p + j\omega$ and the speed of light is c. The source point is denoted by r_q and the radiation distance is calculated as follows:

$$R = |\mathbf{r}_p - \mathbf{r}_q|$$

$$= r_p - \frac{2\mathbf{r}_p \cdot \mathbf{r}_q}{r_p + \sqrt{r_p^2 + r_q^2 - 2\mathbf{r}_p \cdot \mathbf{r}_q}} + \frac{r_q^2}{r_p + \sqrt{r_p^2 + r_q^2 - 2\mathbf{r}_p \cdot \mathbf{r}_q}} \qquad (7.14)$$

Equation (7.13) describes the radiated field under both near- and far-field conditions.

For a large observation distance (far-field) the approximations

$$R = \sqrt{r_p^2 + r_q^2 - 2\mathbf{r}_p \cdot \mathbf{r}_q} \approx r_p - 2\mathbf{r}_p \cdot \mathbf{r}_q \qquad (7.15)$$

and

$$\frac{sR + c}{cR^2} \approx \frac{s}{cr_p} \qquad (7.16)$$

can be used to simplify Equation (7.13):

$$\mathbf{E}_{\text{far}}(\mathbf{r}_p, s) = \frac{1}{2\pi cr_p} e^{-(s/c)r_p} \int_{A_a} [(\mathbf{e}_z \times \mathbf{E}_a(\mathbf{r}_q, s)) \times \mathbf{e}_p] e^{2(s/c)\mathbf{r}_p \cdot \mathbf{r}_q} \, dA_a \qquad (7.17)$$

where $\mathbf{e}_p$ is the unit vector of $\mathbf{r}_p$. This far-field formulation for the radiated electrical field strength consists of two independent terms. The first one describes wave propagation (phase shift and amplitude attenuation) as a function of observation distance. The second term, the integral over the aperture area, is a function of the field distribution inside the aperture and the direction angle. In the far-field formulation just the phase (or time delay) and the attenuation of the radiated field are functions of the observation distance r_p. As a result, the amplitude of the radiated field can be scaled by changing the distance if the far-field approximation is applicable. Because an error less than 5% is acceptable for normal applications, the recent parameter of interest is the observation distance at which the errors caused by the approximations of Equations (7.15) and (7.16) are smaller than this limit.

Considering the far-field approximations, let us assess the approximation errors in the frequency domain first. The approximation (7.15) yields a relative phase shift of

$$\Delta\varphi_{\text{rel}} \leq \frac{r_n^2}{1 + \sqrt{1 + r_n^2}} \qquad (7.18)$$

with the normalized source distance $r_n = r_q/r_p$. It is simple to predict that, if $r_n \leq 0.3$ ($r_p \geq 3r_q$), the relative phase shift will be smaller than 5%.

The amplitude error resulting from the approximation (7.16) can be calculated by

$$\Delta_{\text{rel}} = \frac{\lambda}{r_p 2\pi(1 + r_n^2)} + \frac{1 - \sqrt{1 + r_n^2}}{\sqrt{1 + r_n^2}} \tag{7.19}$$

Using the determined ratio between the observation distance and the source point, the amplitude error will be smaller than 5% if $r_p \geq 3\lambda$. Note that for the case when $r_n \leq 0.1$ and $r_p \geq 12\lambda$ the approximation error will be less than 1%.

7.2.1 Dispersion due to near-field effects

After the well-known frequency domain limits of far-field approximations have been expressed, the behavior of fast transient radiation can be investigated. Assuming

$$\frac{\partial E_a}{\partial t} \gg \frac{E_a c}{r_p}$$

the time domain expression of Equation (7.13) is derived and

$$\mathbf{E}_s(\mathbf{r}_p, t) = \frac{1}{2\pi c r_p} \int_{A_a} \left[\left(\mathbf{e}_z \times \frac{\partial}{\partial t} \mathbf{E}_a \left(t + \frac{\mathbf{r}_p \cdot \mathbf{r}_q}{c} - \frac{r_p}{c} \right) \right) \times \mathbf{r}_p \right]$$

$$* \ \delta \left(t + \frac{\mathbf{e}_p \cdot \mathbf{r}_q}{c} \frac{1 - \sqrt{1 + r_n^2 - 2\mathbf{e}_p \cdot \mathbf{e}_q r_n}}{1 + \sqrt{1 + r_n^2 - 2\mathbf{e}_p \cdot \mathbf{e}_q r_n}} \right.$$

$$\left. - \frac{r_q}{c} \frac{r_n}{1 + \sqrt{1 + r_n^2 - 2\mathbf{e}_p \cdot \mathbf{e}_q r_n}} \right) dA_a \tag{7.20}$$

where $\mathbf{e}_q$ is the unit vector of $\mathbf{r}_q$, '$*$' describes the convolution with respect to time, and $\delta(\cdot)$ is the Dirac delta function. The convolution with the delayed Dirac delta function yields a wider pulse in the near and intermediate field. This near-field dispersion can be ignored if the delay

$$\Delta t \leq \frac{r_q}{c} \frac{r_n}{1 + \sqrt{1 + r_n^2}} \tag{7.21}$$

between the shortest path and the length of a path from an arbitrary point on the aperture is short compared with the rise time of the radiated signal. Using the frequency domain far-field definition we get a delay of $\Delta t = 0.15 r_q/c$. For fast transient signals and large-aperture cross-sections, in particular, this time delay is not acceptable. For a distance-independent pulse shape of transient radiation it is necessary that Δt is small compared with the rise time of the radiated signal. For practical applications the rise time should be six times larger than the longest time delay ($\tau_r \geq 6\Delta t$). Based on this, the transient far-field is given by

$$r_p \geq \frac{3r_q^2}{c\tau_r} = \frac{r_q}{c} \cdot \frac{3r_q}{\tau_r} \tag{7.22}$$

Unlike the frequency domain equation, in this time domain equation the time r_p/c which the wave needs to cross the aperture is weighted with the rise time as a measure of the shortest change in the shape of the signal. By comparing the phase shift with the phase of one oscillation of a continuous wave signal we can get a similar relation in the frequency domain.

7.3 SUITABILITY OF CONVENTIONAL ANTENNAS FOR THE UWB SYSTEM

Antennas can be classified as either *resonant* or *nonresonant*, depending on their design. In a resonant antenna, if the antenna works at its resonance frequency, almost all of the radio signal fed to the antenna is radiated. However, if the antenna is fed with a frequency other than a resonant one, a large portion of the signal will not be radiated. In the case of a resonant antenna, if the frequency range is very wide, a separate antenna must be made for each frequency. On the other hand, a nonresonant antenna can cover a wide range of frequencies. However, special care must be taken when designing the antenna to make it efficient. Moreover, the physical size of currently available nonresonant antennas is not appropriate for portable UWB devices.

It is very difficult to compare a UWB antenna with a conventional antenna, because the traditional performance considerations are based upon continuous wave or narrowband theory. Basic concepts should be kept in mind when a conventional approach is used for UWB technology.

The goal of the UWB antenna designer is to design an antenna with a small size and a simple structure that can produce low distortions, but can provide large bandwidth and omni-directional patterns [80].

7.3.1 Resonant antennas

The most common and easiest antennas for communications are wire antennas. They are some of the cheapest, simplest, and most flexible antennas for many applications. They can be made of a very thin wire, a thicker wire, or a cylinder.

One of the simplest practical antennas is the *dipole antenna* shown in Figure 7.3. A Hertzian electric dipole is shown in Figure 7.4 as a pair of electric charges that vary sinusoidally with time in such a way that at any instant the two charges have equal magnitude but opposite sign.

The Hertzian dipole is an inefficient radiator due to its need for a high voltage to produce a large current. However, this high voltage does not contribute to the radiated power. The resonant dipole (Figure 7.3) was the solution to the inefficiency of the Hertzian dipole (Figure 7.4).

Propagation time is an important parameter in radiating elements. Thus, if an alternating current is flowing in the radiating element as shown in Figure 7.5, the

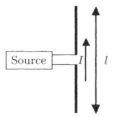

Fig. 7.3   A dipole antenna connected to a source.

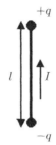

Fig. 7.4   A Hertzian electric dipole.

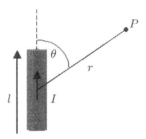

Fig. 7.5   Radiating element.

effect of the current is not immediately noticeable at point P, but only after the period of time necessary to propagate over the distance r.

Current I is the time-varying current and at a distance of $r = 0$ is equal to

$$I = I_0 \sin(\omega t) \tag{7.23}$$

where ω is the radian frequency and t is time. From the Lorentz equations we can introduce the time of propagation. Therefore, Equation (7.23) can be rewritten as

$$I_p = I_0 \sin\left[\omega\left(t - \frac{r}{c}\right)\right] \tag{7.24}$$

where c is the speed of light. With respect to Figure 7.5, Equation (7.24) shows that the current I_p at point P with distance r from the radiating element at time t occurs at the earlier time of $(t - r/c)$.

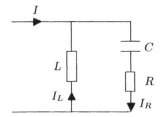

Fig. 7.6 RLC circuit of resonant antenna.

To get a clear understanding of resonant antennas, let us apply a sinusoidal current to the RLC circuit as shown in Figure 7.6. At resonance we will have $\omega^2 LC = 1$, which gives the resonant frequency ω_0 as follows:

$$\omega_0 = \frac{1}{\sqrt{LC}} \tag{7.25}$$

where L is the inductance and C is the capacitance. At this frequency the current in the inductance I_L can be written as

$$I_L = I_0 \frac{Z}{R} \left(\cos \omega_0 t - \frac{R}{Z} \sin \omega_0 t \right) \tag{7.26}$$

where R is the resistance and Z is the impedance equal to

$$Z = \sqrt{\frac{L}{C}} = \frac{1}{C\omega_0} \tag{7.27}$$

Thus, Equation (7.26) can be rewritten as

$$I_L = I \left(\cos \omega_0 t - \frac{R}{Z} \sin \omega_0 t \right) \tag{7.28}$$

where

$$I = I_0 \frac{Z}{R} = \frac{I_0}{RC\omega_0} \tag{7.29}$$

If R is sufficiently small so that it can be neglected, the term $(R/Z)\sin \omega_0 t$ will be zero. Therefore, Equation (7.28) can be rewritten as

$$I_L = I \cos \omega_0 t \tag{7.30}$$

The current I_L shows the behavior of a resonant antenna. The principle of resonance in a resonant antenna has been used to increase the current. If an UWB impulse is fed to this kind of antenna a *ringing* effect will occur. This severely distorts the pulse, spreading it out in time.

Another reason that dipole antennas are not suitable for the UWB system is due to the standing wave produced by the reflection from the end points of the antenna. This is shown graphically in Figure 7.7.

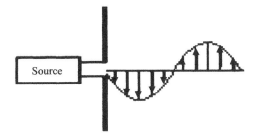

Fig. 7.7 Electromagnetic field and standing wave generated by a dipole antenna.

7.3.2 Nonresonant antennas

An antenna designed to have approximately constant input impedance and radiation characteristics over a wide range of frequencies is called a *nonresonant* antenna, or a *frequency-independent* antenna.

In a nonresonant antenna the maximum dimension (size) will be set by the limit of the lowest frequency to be radiated. The higher frequency limit will be decided by how precisely the input terminal region can be constructed. In fact, a true frequency-independent antenna is only a theoretical construct. In practice, frequency independence is only over a limited frequency bandwidth.

If the impedance and radiation characteristics of the antenna do not change significantly over about an octave or more, the antenna is called a *wideband antenna*.

From sinusoidal wave antenna theory we find that there are many types of antennas which can propagate nonsinusoidal waves. The *log-periodic* antenna and *spiral* antenna are examples of wideband antennas. However, these antennas are likely to be dispersive and inappropriate for very short pulses such as UWB signals. They radiate different frequency components from different parts of the antenna. Therefore, the radiated waveform will be both extended and distorted.

In essence, a tradeoff is necessary to obtain an efficient, electrically small antenna which is suitable for the required application.

7.3.3 Difficulties with UWB antenna design

In UWB systems, antennas are significant pulse-shaping filters. Any distortion of the signal in the frequency domain (which is a filtering operation) causes distortion of the transmitted pulse shape, therefore increasing the complexity of the detection mechanism at the receiver.

UWB antennas require the phase center and the VSWR to be constant across the whole bandwidth of operation. A change in phase center may cause distortion of the transmitted pulse and worse performance at the receiver. Therefore, the design of antennas for UWB signal radiation is one of the main challenges for the UWB system, especially when low-cost, geometrically small, and radio-efficient structures are required for typical consumer communication applications.

Conventional antennas are designed to radiate only over the relatively narrow range of frequencies used in conventional narrowband systems. If an impulse is fed to such an antenna, it tends to ring, which severely distorts the pulse and spreads it out in time. We have shown that resonant antennas are unsuitable for the UWB system, as they can only radiate sinusoidal waves at the resonance frequency.

On the other hand, to make an effective nonresonant antenna is not an easy task. One way to make a UWB antenna is to put the antenna resonating frequency above the UWB band. However, as the physical dimensions of antennas decrease, antennas will lose efficiency. Another way is to build an antenna with a lower quality factor, which results in a wider bandwidth but with lower efficiency.

The challenge of wide bandwidth antennas is well understood in other applications; however, work on wide bandwidth antennas for UWB applications is ongoing and far from complete.

7.4   IMPULSE ANTENNAS

A few high-quality, laboratory-grade, nondispersive UWB antennas are commercially available, although they targeted at laboratory usage rather than commercial consumer products. For example, Farr Research, Inc. [81], offers several antennas that operate across several decades of bandwidth. However, the high price range of these antennas make them less suitable for most commercial applications and infeasible for portable or handheld applications. There is a great need for a low-cost, easy-to-manufacture antenna that is omni-directional, radiation-efficient, and has a stable UWB response.

7.4.1   Conical antenna

The conical antenna (Figure 7.8) suspended over a large metal ground plane is the preferred antenna for transmitting known transient electromagnetic waves. This type of antenna is used as a reference transient transmitting antenna. It radiates an electromagnetic (E-M) field that is a perfect replica of the driving point voltage waveform.

The conical antenna can be analyzed as one-half of the bi-conical transmission line [82] with uniform characteristic impedance given by

$$Z_{\text{cone}} = \left(\frac{\eta}{2\pi}\right) \ln\left[\cot\left(\frac{\theta_0}{2}\right)\right] \qquad (7.31)$$

and where η is the free space impedance (377 Ω) and θ_0 is the solid half-angle of the cone. A 4° cone has an impedance of 200 Ω, while a 47° cone is required to achieve an impedance of 50 Ω.

The electric field (E field) generated at radius r and angle θ is given by

$$E_\theta(r,t) = \frac{V_{\text{base}}(t - r/c)}{r\sin(\theta)\ln[\cot(\theta_0/2)]} \qquad (7.32)$$

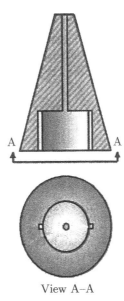

View A–A

Fig. 7.8 Conical antenna.

for $t < l/c$, where l is the length of the antenna.

The driving point voltage at the base of the conical antenna is given by

$$V_{base}(t) = V_{gen}(t)\frac{R_{cone}}{R_{cone} + R_{gen}} \tag{7.33}$$

where $V_{gen}(t)$ is the source voltage, R_{cone} is the cone antenna resistance, and R_{gen} is the source resistance.

Equation (7.32) is only valid for the time window up to $T_w < l/c$. For $t > T_w$ the original incident wave launched on the antenna has reached the end of the antenna and is reflected, thus setting up a series of multiple reflections on the antenna's radiating element. The radiated E field is thus no longer a replica of the driving point voltage, but is corrupted by multiple reflections. If a conical antenna is used as a receiving antenna, its output is the *integral* of the incident E field.

7.4.2 Monopole antenna

The monopole antenna is sometimes used as a simpler version of the conical antenna for transmitting UWB signals that are similar in wave shape to the driving point voltage. However, its radiated fields are not as uniform as those of the conical antenna. Its driving point impedance is not constant, but increases as a function of time. This leads to distortion of radiated E-M fields.

Time domain studies of a monopole antenna confirm this statement. When a monopole is used for receiving transient E-M fields, its output is the *integral* of the incident E field. The integration effect can be explained simply. Assume an impulsive

E field is incident upon the monopole at a 90° angle to its axis. Thus, a current is induced simultaneously in each differential element dx of the antenna. These current elements di thus start to flow toward the output connector of the antenna. They do not arrive at the output simultaneously but in sequence. Thus, the output appears as a step function, which is the integral of the incident impulse. The integrating effect of this antenna only lasts for $t < l/c$.

7.4.3   D-dot probe antenna

The D-dot antenna is basically an extremely short, monopole antenna. The equivalent antenna circuit consists of a series capacitance and a voltage generator $V_a(t)$ as follows:

$$V_a(t) = h_a E_i(t) \qquad (7.34)$$

where h_a is the height of the antenna and $E_i(t)$ is the E field. For a very short monopole, antenna capacitance is very small and the capacitor thus acts like a differentiator to transient E-M fields. Therefore, the output from a D-dot probe antenna is the *derivative* of the incident E field.

To determine the actual wave shape of the incident E field we must integrate the output from the D-dot probe. When the frequencies become too high, the D-dot probe loses its derivative properties and it becomes a monopole antenna. This happens when the length of the D-dot probe approaches a quarter wavelength of the incident wave.

7.4.4   TEM horn antenna

Transient electromagnetic (TEM) horns are the most effective receiving antennas for making direct measurement of transient E-M fields (Figure 7.9). The TEM horn antenna is basically an open-ended, parallel-plate transmission line. It is typically built using a taper from a large aperture at the receiving input down to a small aperture at the coax connector output. The height-to-width ratio of the parallel plate is maintained constant along the length of the antenna to maintain a uniform characteristic impedance. The output from a TEM antenna is

$$V_{\text{out}}(t) = \frac{h_{\text{eff}} E_i(t) R_{\text{load}}}{R_{\text{ant}} + R_{\text{load}}} \qquad (7.35)$$

where h_{eff} is the effective height at the aperture of the antenna, and R_{load} and R_{ant} are load and antenna resistances, respectively.

Ideally, to suppress multiple reflections, antenna impedance should match the 50 Ω output cable. However, to optimize sensitivity most TEM antennas are designed with an antenna impedance of 100 Ω. If a TEM antenna is used as a transmit antenna the radiated E field is the *derivative* of the input driving point voltage. We can see that the TEM horn antenna is capable of radiating and receiving very fast, but still clean, pulses, which is of course important for many applications. The cleaner the pulse, the cleaner the backscattered signal, and the easier it will be to post-process and interpret the data.

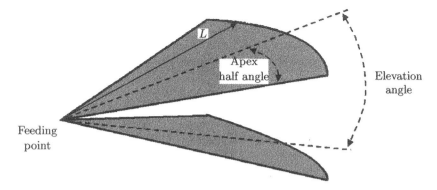

Fig. 7.9 Typical TEM horn antenna with length L.

7.4.5 Small-size UWB antenna

In order to ensure design flexibility and optimal performance in devices such as wireless USB and devices enabled by it, such as digital cameras, video recorders, PC peripherals, PDAs, mobile phones, wireless compact flash and secure digital (SD) cards, and other consumer electronic devices, small surface-mounted monopole chip antenna have been designed and implemented for UWB applications.

As an example, Fractus UWB chip antenna [83] is a high-performance, cost-effective antenna designed to meet the requirements of reference designers considering the MultiBand OFDM Alliance (MBOA) recommendations for UWB devices.

Table 7.1 demonstrates the parameters of the Fractus UWB chip antenna.

Table 7.1 Parameters of the Fractus UWB chip antenna.

Parameter	Value
Frequency range	3.1–5 GHz
Efficiency	>60%
Peak gain	<4 dBi
Gain flatness	<±2 dB
VSWR	<2:1
Weight	0.20 g
Temperature	−40 to +85°C
Impedance	50 Ω unbalanced
Dimensions	$10 \times 10 \times 0.8$ mm^3

7.4.6   Conclusion

Most practical UWB systems in fact will *not* use the antennas discussed above. To meet FCC spectrum requirements, considerable band-pass filtering will need to be done.

As a conclusion we can say that, although the classical antennas have proven their use in different applications, none of them can be used for UWB applications and there is a need to look for new types of antennas. To do so we can summarize the following technical and practical design goals for the UWB antenna:

- The antenna must be able to radiate or receive fast electromagnetic transients with frequencies between 3.1 GHz and 10.6 GHz.

- We want an antenna which can be used off ground, not only for safety reasons but also to improve the mobility of the sensor.

- Another criterion to guarantee high mobility is the dimensions and weight of the antenna. Small antennas are also better for handheld applications.

- The antennas must be cheap to produce.

7.5   BEAMFORMING FOR UWB SIGNALS

Array signal processing or beamforming involves the manipulation of signals induced on the elements of an array system or transmitted from the elements and received at a distant point from the array. In a conventional narrowband beamformer the signals corresponding to each sensor element are multiplied by a complex weight to form the array output. As the signal bandwidth increases, the performance of the narrowband beamformer starts to deteriorate because the phase provided for each element and the desired angle is not a function of frequency and, hence, will change for the different frequency components of the communication wave.

For processing broadband signals a tapped delay line (TDL) can normally be used on each branch of the array. The TDL allows each element to have a phase response that varies with frequency, compensating for the fact that lower frequency signal components have less phase shift for a given propagation distance, whereas higher frequency signal components have greater phase shift as they travel the same length. This structure can be considered to be an equalizer that makes the response of the array the same across different frequencies. In addition to UWB signals, inherent baseband signals, such as audio and seismic signals, are examples of wideband signals. In sensor array processing applications, such as SS communications and passive sonar, there is also growing interest in the analysis of broadband sources and data.

7.5.1 Basic concepts

Most of the smart antennas proposed in the literature are narrowband beamformers. The antenna spacing of narrowband arrays is usually half of the wavelength of the incoming signal which is assumed to have a fractional bandwidth of less than 1%.

By definition the fractional bandwidth of a signal is the ratio of the bandwidth to the center frequency (Equation (7.11)). Wideband arrays are designed for a fractional bandwidth of up to 20% and UWB arrays are proposed for a fractional bandwidth of 20 to 200%. Wideband and UWB arrays use the same antenna spacing for all frequency components of the arriving signals. Usually, the inter-element distance d is determined by the highest frequency of the input wave. In a uniform, one-dimensional linear array we can write

$$d = \frac{c}{2f_h} \tag{7.36}$$

Wideband and UWB antenna arrays use a combination of spatial filtering and temporal filtering. On each branch of the array a filter allows each element to have a phase response that varies with frequency. As a result, the phase shifts due to higher and lower frequencies are equalized by temporal signal processing.

Figure 7.10 shows the general structure of a wideband array antenna system in *receiving mode*. The TDL network permits adjustment of gain and phase as desired at a number of frequencies over the band of interest. The far-field wideband signal is received by N antenna elements. Each element is connected to $M - 1$ delay lines, with the time delay of T seconds. The delayed input signal of each element is then multiplied by a real weight C_{nm}, where $1 \leq n \leq N$ and $1 \leq m \leq M$. If the input signals are denoted by $x_1(t), x_2(t), \ldots, x_N(t)$ the output signal which is the sum of all intermediate signals can be written as

$$y(t) = \sum_{n=1}^{N} \sum_{m=1}^{M} C_{nm} x_n \left(t - (m-1)T \right) \tag{7.37}$$

In a linear array, such as in Figure 7.10 [84], the signals $x_n(t)$ are related according to the angle of arrival and the distance between elements. Figure 7.11 shows that a time delay τ_n exists between the signal received at element n and at reference element $n = 1$. This amount of delay can be found as follows:

$$\tau_n = (n-1)\frac{d}{c} \sin \theta \tag{7.38}$$

where d is inter-element spacing and c is propagation speed. It is assumed that the incoming signal is spatially concentrated around the angle θ. Using the time delays corresponding to the antenna elements, we can now write $x_n(t)$ with respect to $x_1(t)$ as follows:

$$x_n(t) = x_1(t - \tau_n)$$
$$= x_1 \left(t - (n-1)\frac{d}{c} \sin \theta \right) \tag{7.39}$$

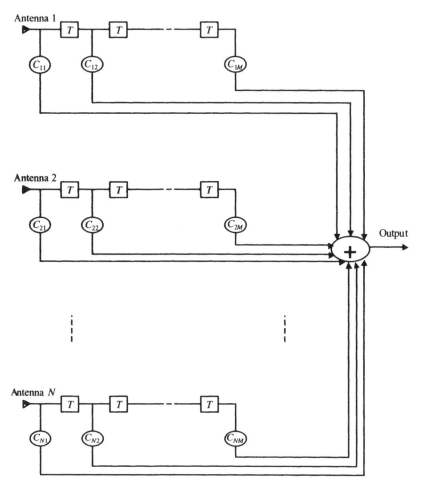

Fig. 7.10 General structure of a TDL wideband array antenna using N antenna elements and M taps.

Different methods and structures can be utilized for computation of the adjustable weights of a wideband beamforming network. In the following section we will explain one of them.

7.5.2 A simple delay-line transmitter wideband array

The problem of designing a uniformly spaced array of sensors for operation at a narrowband frequency domain is well understood. However, when a single frequency design is used over a wide bandwidth the array performance degrades significantly. At lower frequencies the beamwidth increases, resulting in reduced spatial resolution; at frequencies above the narrowband frequency the beamwidth decreases and grating lobes may be introduced into the array beam pattern.

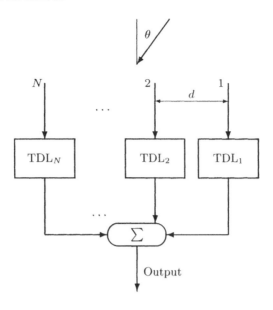

Fig. 7.11   The incoming signal arrives at the antenna array with angle θ.

Figure 7.12 shows the basic structure of a delay-line wideband transmitter array system. The adjustable delays T_n, $n = 1, 2, \ldots, N$ where N is the number of antenna, are controlled by the desired angle of the main lobe of the directional beam pattern θ_0 as follows:

$$T_n = T_0 + (n - 1)\frac{d}{c} \sin \theta_0 \qquad (7.40)$$

where d is the inter-element spacing. The constant delay $T_0 \geq (N - 1)d/c$ is required because without T_0 a negative delay will be obtained for negative values of θ_0 and this cannot be implemented. The signal received at the far-field in the direction of θ $(-90° < \theta < +90°)$ is equal to

$$y(t) = A(\theta) \sum_{n=1}^{N} x_n(t - \tau_n)$$

$$= A(\theta) \sum_{n=1}^{N} x(t - T_n - \tau_n) \qquad (7.41)$$

where $x_n(t)$ indicates the transmitted signal from transducer n, τ_n is the delay due to the different distances between the elements and the receiver, and $A(\theta)$ is the overall gain of the elements and the path. The time delay τ_n regarding Figure 7.12 is equal to

$$\tau_n = \tau_0 - (n - 1)\frac{d}{c} \sin \theta \qquad (7.42)$$

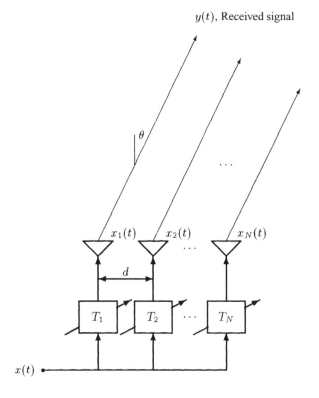

Fig. 7.12 Beam formation using adjustable delay lines.

where τ_0 is the constant transmission delay of the first element, which is independent of θ. The gain $A(\theta)$ can be decomposed into two components as follows:

$$A(\theta) = A_1(\theta)A_2 \tag{7.43}$$

where $A_1(\theta)$ is the angle-dependent gain of the elements and A_2 is attenuation due to the distance. Substituting Equations (7.42) and (7.43) into Equation (7.41) yields

$$y(t) = A_1(\theta)A_2 \sum_{n=1}^{N} x\left(t - \alpha_0 - (n-1)\frac{d}{c}(\sin\theta_0 - \sin\theta)\right) \tag{7.44}$$

where $\alpha_0 = T_0 + \tau_0$. In the frequency domain we may write

$$H(f,\theta) = \frac{Y(f,\theta)}{X(f)}$$

$$= A_1(\theta)A_2 \exp[-j2\pi f\alpha_0] \sum_{n=1}^{N} \exp[-j2\pi f(n-1)(d/c)(\sin\theta_0 - \sin\theta)]$$

$$= A_1(\theta)A_2 \exp[-j2\pi f\alpha_0] \exp[-j\pi f(N-1)(d/c)(\sin\theta_0 - \sin\theta)]$$
$$\times \frac{\sin[\pi f N(d/c)(\sin\theta_0 - \sin\theta)]}{\sin[\pi f(d/c)(\sin\theta_0 - \sin\theta)]} \qquad (7.45)$$

From this equation we can derive several properties of a wideband delay beamformer. An important characteristic of the beamformer is the directional patterns for different frequencies.

Example 7.1

Consider a UWB signal with a center frequency of 6 GHz and a bandwidth of 2 GHz. Calculate and sketch the normalized amplitude of Equation (7.45) for $\theta_0 = 10°$, $-90° < \theta < +90°$, $N = 10$, $d = 2.14$ cm, $c = 3 \times 10^8$ m/s, and perfect antennas (i.e. $A(\theta) = A_2$).

Solution

The result is plotted in Figure 7.13. We observe that at $\theta = \theta_0$ the frequency independence is perfect, but as we move away from this angle the dependence increases. Nevertheless, the beamformer is considered wideband with a fractional bandwidth of two-sixths or 33%.

Increasing the inter-element spacing has positive and negative consequences. As we will shortly see, it will produce a sharper beam and it is clearly more practical. On the other hand, this increase will result in some extra main lobes in the same region of interest (i.e. $-90° < \theta < +90°$).

7.5.2.1 Angles of grating lobes We now derive the angles of grating lobes and conditions for their existence. Assuming perfect antennas (i.e. $A(\theta) = A_2$), we can write from Equation (7.45), for $\theta = \theta_0$,

$$|H(f, \theta_0)| = A_2 N \qquad (7.46)$$

This situation can happen for some other angles, denoted by θ_g. To calculate θ_g, it follows from Equation (7.45) that

$$|H(f, \theta_g)| = A_2 N = A_2 \frac{\sin[\pi f N(d/c)(\sin\theta_0 - \sin\theta_g)]}{\sin[\pi f(d/c)(\sin\theta_0 - \sin\theta_g)]} \qquad (7.47)$$

Now, Equation (7.47) should be solved for θ_g:

$$\sin\left[\pi f \frac{d}{c}(\sin\theta_0 - \sin\theta_g)\right] = 0$$

or

$$\pi f \frac{d}{c}(\sin\theta_0 - \sin\theta_g) = m\pi \qquad (7.48)$$

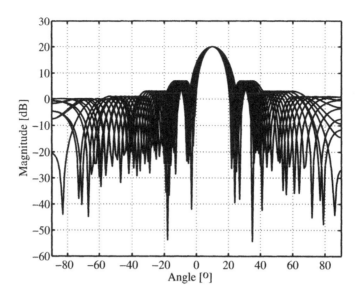

Fig. 7.13 Directional patterns of a delay beamformer for 11 frequencies uniformly distributed from 5 to 7 GHz.

where $m = \pm 1, \pm 2, \ldots$. The result is

$$\theta_g = \sin^{-1}\left(\sin\theta_0 - m\frac{c}{fd}\right) \tag{7.49}$$

The first grating lobes are given for $m = \pm 1$. The necessary condition for having no grating lobe for a beamformer is that θ_g does not exist for any values of $-90° < \theta_0 < +90°$. The worst case happens for $\theta_0 = \pm 90°$ and the condition of no grating lobe can be inferred from Equation (7.49) as

$$\frac{c}{fd} \geq 2 \quad \text{or} \quad d \leq \frac{c}{2f} = \frac{\lambda}{2} \tag{7.50}$$

where λ indicates the wavelength. It is interesting to note that θ_g is not a function of N, but is very dependent on d. To show this more adequately, Figure 7.13 is replotted for $d = 8.57$ cm in Figure 7.14. We observe that the nearest grating lobes for $f = 6$ GHz are at 49.2° and −24.2° and are in agreement with Equation (7.49). The frequency dependence of the beam patterns increases as we move away from the desired angle.

7.5.2.2 Inter-null beamwidth   Comparing Figures 7.13 and 7.14 reveals that the main beamwidth of Figure 7.14 is less than that of Figure 7.13. The corresponding equation can be derived easily. The inter-null beamwidth (INBW) is defined as the difference between the nearest two nulls around the desired angle. Starting from Equation (7.45)

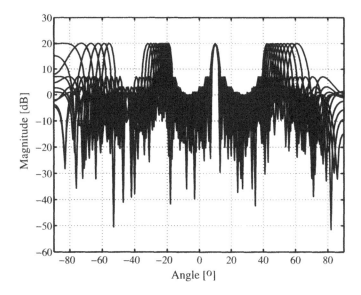

Fig. 7.14 Grating lobes appear as a result of the increase of spacing between antennas.

and equating it to zero gives the following:

$$\pi f N \frac{d}{c}(\sin\theta_0 - \sin\theta) = m\pi \tag{7.51}$$

where $m = \pm 1, \pm 2, \ldots$ The first two angles around θ_0 are denoted by θ_1 and θ_2 and are computed from Equation (7.51) for $m = +1$ and $m = -1$, respectively:

$$\theta_1 = \sin^{-1}\left(\sin\theta_0 - \frac{c}{f d N}\right) \tag{7.52}$$

$$\theta_2 = \sin^{-1}\left(\sin\theta_0 + \frac{c}{f d N}\right) \tag{7.53}$$

Hence, the INBW, $\Delta\theta = \theta_2 - \theta_1$, is written as

$$\text{INBW} = \sin^{-1}\left(\sin\theta_0 + \frac{c}{f d N}\right) - \sin^{-1}\left(\sin\theta_0 - \frac{c}{f d N}\right) \tag{7.54}$$

It is clear that for $|\sin\theta_0 \pm c/f d N| > 1$ there exists no null on the left-hand or right-hand side of the main angle θ_0. As a special case, for $\theta_0 = 0$ we have

$$\text{INBW}_0 = 2\sin^{-1}\left(\frac{c}{f d N}\right) \tag{7.55}$$

that is, increasing d lowers the INBW and produces sharper beams. It is easy to test Equation (7.54) for values of the first and second cases, which are illustrated in Figures 7.13 and 7.14.

As is obvious from Equations (7.54) or (7.55), INBW is a function of frequency f. To observe the effect of frequency variations on the beam pattern of the delay-line beamforming network we repeat Example 7.1 for a wide frequency range from 4 to 8 GHz and an inter-element spacing of 1.87 cm. The results are shown in Figure 7.15 for frequencies of 4, 5, 6, 7, and 8 GHz. The computed values of INBW for these frequencies are 47.3, 37.4, 31, 26.5, and 23.1°, respectively.

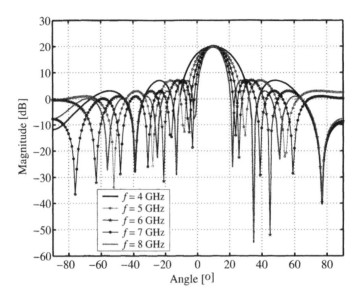

Fig. 7.15 Directional patterns of the delay beamformer for five different frequencies show that the beamwidth is very sensitive to frequency.

From the forgoing discussion we can conclude that pure delay-line wideband antenna arrays have the following properties:

- a relatively simple structure using only a variable delay element;

- no multiplier in the form of amplification or attenuation;

- a perfect frequency independence characteristic only for the desired angle of the array;

- their INBW and side lobe characteristics vary considerably with frequency of operation.

Because of the existence of some distinctive differences between conventional and UWB antenna arrays, the well-known conventional concepts of phased array antennas have to be modified appropriately to accommodate UWB signals. One significant difference from narrowband theory is that frequency domain analysis alone is insufficient to treat UWB arrays. In fact, the time domain may be a more natural setting for understanding and analyzing the radiation of UWB signals.

7.6   RADAR UWB ARRAY SYSTEMS

Advanced radar systems employ array antennas in their passive or active forms to achieve the required total power, high-resolution directive beam pattern, electronic and automatic beam steering, and interference suppression through side lobe nulling [85]. The conventional method of beamforming based on periodic sinusoidal waves results in a beam pattern, or array factor $F(\theta)$, of the form

$$F(\theta) = \frac{\sin(L\pi\theta/\lambda)}{L\pi\theta/\lambda} = \frac{\sin(Lf\pi\theta/c)}{Lf\pi\theta/c} \tag{7.56}$$

where L is the array size and θ is a function of the angle of incidence or radiation. The array factor in Equation (7.56) results in the well-known equation for resolution angle

$$\varepsilon = \frac{k\lambda}{L} = \frac{kc}{Lf} \tag{7.57}$$

where k is a constant usually set equal to one.

Array beamforming based on Gaussian pulses yields the array factor as follows [86]:

$$A(\theta) = \frac{\mathrm{erf}[\sqrt{\pi}L(\Delta f)\theta/2c]}{\sqrt{\pi}L(\Delta f)\theta/2c} \tag{7.58}$$

where $\mathrm{erf}[\cdot]$ is the well-known error function and Δf is the approximate bandwidth of the pulses. If we define the parameter ρ as

$$\rho = \frac{L(\Delta f)}{c} \tag{7.59}$$

the variation of the array factor as a function of θ and ρ can be demonstrated as in Figure 7.16.

As an example, for $\Delta f = 4$ GHz, $c = 3 \times 10^8$ m/s, and $L = 0.3$ m we have $\rho = 5$, and from Figure 7.16 the array factor is equal to 1.128 and 0.141 for $\theta = 0°$ and $\theta = 2°$, respectively. An increase in frequency bandwidth and in the array length (e.g., the number of antenna elements) will increase the parameter ρ. An increase in ρ decreases the array factor at all angles other than zero degrees. In contrast to the beam patterns of Equation (7.56), which include distinguishable side lobes, $A(\theta)$ is a monotonically decreasing function of angle θ.

The resolution angle from the array factor $A(\theta)$ is given by

$$\varepsilon = \frac{Kc}{(\Delta f)L} = \frac{K(\Delta T)c}{L} \tag{7.60}$$

where K is a constant usually set equal to one. The resolution angle for nonsinusoidal signals in Equation (7.60) is a function of the array size L and the bandwidth Δf. An increase in bandwidth results in simultaneous improvement in range resolution and angular resolution. This feature is of interest for UWB high-resolution imaging radar.

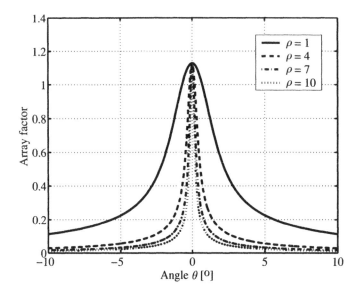

Fig. 7.16 Array factor for beamforming based on Gaussian pulses as a function of angle θ and ρ.

7.7 SUMMARY

In this chapter the basic knowledge regarding UWB antennas and arrays was presented. First, the fundamental equations governing antenna theory were mentioned and important parameters involved in antenna analysis were explained. We understood that the radiation of short-duration UWB signals from an antenna is significantly different from long-duration narrowband signals. By studying resonant and nonresonant antennas we showed that conventional antennas are not suitable for the UWB system.

Different antenna elements applicable to UWB systems, such as the conical antennas, monopole antennas, D-dot probe antennas, and TEM horn antennas, were introduced. We concluded that, although the classical antennas have proven their use, none of them are exactly suitable for UWB applications, and, therefore, there is a need to look for new types of antennas.

As a conventional and straightforward method for UWB beamforming we studied the delay-line beamformer. Several parameters of this array, such as the beamwidth and angles of grating lobes, were derived and discussed. Finally, we introduced radar UWB array systems and several factors affecting their performance.

Problems

Problem 1. Derive the radiation distance as given in Equation (7.14).

Problem 2. Using the *RLC model* for resonance antenna given in Figure 7.6 verify the validity of Equation (7.26).

Problem 3. Discuss the difficulties of UWB antennas that are appropriate for communication systems with respect to the following:

(i) size;

(ii) cost;

(iii) VSWR;

(iv) frequency response; and

(v) on chip design.

Problem 4. You can find several algorithms for wideband smart antennas in the recent literature. Investigate one of them and discuss the possible applications of that technique for UWB systems.

Problem 5. In Example 7.1 consider

$$d = \frac{4c}{f_h} \qquad (7.61)$$

(i) Find d.

(ii) Calculate and sketch the INBW as a function of frequency f, $f_l < f < f_h$.

(iii) Compute the first two grating lobes at each side of the main lobe for $f = 5$, 6, 7 GHz.

(iv) Sketch the beam patterns of the array for 21 frequencies equally distanced from 5 to 7 GHz.

Problem 6. Derive Equation (7.56).

Problem 7. Referring to [86], derive the formula for the array factor used in array beamforming based on the Gaussian pulses given in Equation (7.58).

8

Position and location with UWB signals

Wireless UWB positioning techniques can provide real-time indoor and outdoor precision tracking for many applications. Some potential uses include locator beacons for emergency services and mobile inventory, personnel and asset tracking for increased safety and security, and precision navigation capabilities for vehicles and industrial and agricultural equipment. The characteristics of UWB signals provide the potential of highly accurate position and location estimation.

In this chapter we explain the fundamentals of positioning and location using UWB signals and systems. Aspects of this topic, such as resolution and timing issues in a practical environment, are considered.

8.1 WIRELESS POSITIONING AND LOCATION

Positioning is defined as the 'determination of the location of somebody or something'. The use of electronic distance measurement techniques to position humans or objects derived from hyperbolic aircraft navigation systems was first developed during World War II. A variety of systems have been used since that time, most of which became quickly obsolete when the GPS became fully operational. However, basic operating concepts have not changed significantly.

This section describes wireless distance measurement and positioning principles. Land-based or terrestrial positioning systems are distinguished from satellite or extraterrestrial positioning systems. All these systems use time difference and trilateration techniques to estimate the position.

Ultra Wideband Signals and Systems in Communication Engineering   Second Edition
M. Ghavami, L. B. Michael and R. Kohno   © 2007 John Wiley & Sons, Ltd

8.1.1 Types of wireless positioning systems

A good method of classifying wireless positioning systems is by their operating frequencies. The frequency generally determines the operating range and accuracy and, in turn, a system's suitability for a particular application. In general, the higher the frequency of the electronic positioning system, the more accurate the resultant position becomes. Systems in the medium frequency range and below are typically hyperbolic phase/pulse differencing and can reach far beyond the visible or microwave horizons. These systems are more suited for long-range navigation purposes. Wireless indoor tracking systems that locate people and objects use high frequency and bandwidth radio signals.

8.1.1.1 Low-frequency positioning systems Low-frequency time-differencing positioning systems are suitable only for long-range navigation problems. Daily calibration is critical if absolute accuracy is to be maintained.

8.1.1.2 Medium-frequency positioning systems The first medium-frequency positioning systems were deployed in the mid-1950s and were used up to the early 1970s (they are no longer used today). Systems in this frequency range operated by time- or phase-differentiating methods and required repeated calibration and continual monitoring.

8.1.1.3 Super-high-frequency positioning systems Microwave and UWB systems in the super-high-frequency (SHF) range are most commonly used over relatively limited distances up to 100 m and can provide the highest distance accuracy measurements.

8.1.2 Wireless distance measurement

Most wireless distance measurement systems operate either by resolving two-way phase delays of a modulated electromagnetic carrier monopulse signal between the object and the reference transmitter or by measuring the two-way propagation time of a coded electromagnetic pulse between these points. On the other hand, GPS operates in a similar manner to conventional systems, except that propagation distances from the satellites are one way. Microwave pulsing systems measure the round-trip propagation delay of a pulse.

For a pulsing system the round-trip distance is computed by multiplying the measured elapsed delay, taking account of the internal system time delays, by the assumed velocity of propagation of electromagnetic energy. The distance, or *range*, is computed by the equation

$$d = c\frac{t_m - t_d}{2} \tag{8.1}$$

where c is the assumed velocity of propagation [m/s], t_m is the measured round trip time delay [s], and t_d is the summation of internal system delays [s]. The resultant distance d will be in meters.

8.1.2.1 Distance determination Under ideal circumstances and with repeated measurements the time delay t_m can be measured fairly accurately and far more accurately when modulated phase comparison techniques are employed, such as with infrared and some microwave and UWB systems. However, at least two factors on the right-hand side of Equation (8.1) are subject to both random and systematic errors. The only way to minimize these errors is by external and internal calibration of the equipment.

The whole internal system delay t_d can be controlled effectively on some modern pulsing systems. Such control is often termed *self-calibrating*. Other local anomalies, or inherent system measurement instabilities, cannot be controlled or corrected by the measurement system. Thus, an independent, on-site calibration must be performed if errors due to these sources become significant, which is normally the case. As a result a calibrated microwave positioning system can measure range to an accuracy of between ±3 m and ±10 m (with 95% rms).

8.1.2.2 Velocity of propagation Variations in the velocity of propagation in air are caused by changes in air density due to temperature, humidity, and air pressure. The effect on land-based microwave positioning systems is more pronounced than on light waves. Assumed stability in the pulsing system time $(t_m - t_d)$ or phase measurement process cannot be guaranteed. Periodic, independent calibration is essential to check this stability.

8.1.2.3 Antenna considerations Electromagnetic wave propagation and refraction problems may exist in some areas. Weather, especially humidity and temperature, affects propagation through the air. Unwanted reflections of the signals can be received at the antenna. Directional antennas may be used to boost a signal into an antenna. This is possible by, for example, using a set of antenna elements and a beamforming algorithm. Circular polarization is a technique used to reduce multipath effects. Another technique used is antenna diversity, which switches from one antenna to another to reduce multipath fading effects.

8.1.2.4 Multipath propagation effects Multipath propagation is a major cause of systematic errors. Errors due to this effect are difficult to detect. Consideration of multipath during antenna placement, antenna design, and other internal electronic techniques and filters is required to identify and minimize multipath propagation effects. Antenna spacing or systems with circular polarization are recommended to minimize the possibility of degrading effects.

8.1.3 Microwave positioning systems

These systems were first used in the early 1970s. They effectively replaced medium-frequency positioning methods that had been used since the 1950s. Up until the mid-1990s, microwave positioning systems were the primary positioning systems nearly everywhere. After 1992, when full-coverage differential GPS became available, the use of microwave systems rapidly declined.

Positioning by microwave systems is accomplished by determining the coordinates of the intersection of two or more measured ranges from known control points. This process is termed *trilateration*. When two circular ranges are measured, two intersection points result, one on each side of the fixed baseline connecting the reference stations. The ambiguity is usually obvious and is controlled by either initializing the computing system with a coordinate on the desired side of the baseline or referencing the point relative to the baseline azimuth.

8.1.3.1 Automated tracking   When automated positioning systems were used the range intersection coordinates were automatically computed and transformed relative to the project alignment coordinate system. These data were then applied to an analog or digital course indicator, allowing any particular cross-section or offset range to be tracked.

8.1.3.2 Positioning accuracy   The positional accuracy of an intersection position (Figure 8.1) is a function of the range accuracy and angle of intersection of the ranges. The angle of intersection varies relative to the baseline. Assuming both ranges have equal values, positional accuracy can be estimated from

$$\text{Positional accuracy} = 2.447\sigma \csc(A) \tag{8.2}$$

where σ is the estimated standard error of measured range distance and A is the angle of intersection of two ranges at the destination.

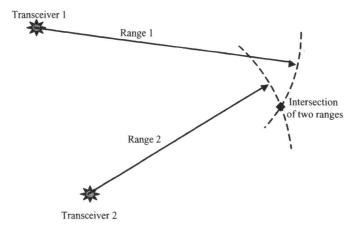

Fig. 8.1   Fundamentals of the range intersection method.

Since A has a major effect on positional accuracy, quality control criteria will restrict surveys within intersection tolerances (e.g., A must be between 45° and 135°). The accuracy of microwave positioning systems is difficult to estimate since it is not constant with distance from a transceiver.

8.1.3.3 Multiple-range positioning technique This method is an expansion of the two-range method described above and was developed in 1979. In this case, three or more ranges are simultaneously observed and positional redundancy results. The position is determined from the computed coordinates of the intersections of three or more range circles. Since each range contains observational errors, all the circles will not intersect at the same point. In the case of three observed ranges, three different coordinates result. Four ranges result in six separate coordinates. The final position is derived by adjustment of these redundant coordinates, usually by a least-squares minimization technique. Some automated microwave positioning systems simply use the strongest angle of intersection as the 'adjusted' position, and others take the unweighted average of all the intersecting coordinates.

Using multiple ranging can minimize positional uncertainties. The coordinated position contains redundancy and can be adjusted. Such a process reduces geometrical constraints and provides an opportunity to evaluate the resultant positional accuracy. This is accomplished by evaluating the best estimate inside the so-called triangle of error which occurs when three or more position lines containing errors intersect. A plot of the simple case of three intersecting ranges is shown in Figure 8.2. The position of the target is obtained by adjusting the three ranges to a best fit.

Assessment of the range measurement accuracy may be obtained by computing the residual range errors v_1, v_2, and v_3 for each position (Figure 8.2). These are the corrections added to each range so that all ranges intersect at the same point. When a least-squares type of adjustment is performed the sum of the squares of the residual errors v_i is minimal. The magnitudes of these residual range corrections provide the statistics for an accuracy estimate of the observed distances or, more practically, an approximate quality control indicator. When a least-squares adjustment is performed, it is possible to obtain an accuracy estimate of the positional rms error. Automated software can provide such data at each position update. If known, different weights may be assigned to individual range observations. This proves useful when different types of positioning systems are mixed.

The residual range errors v_i which result from comparing observed distances with the distances between the adjusted position and remote transmitters could be used to evaluate the accuracy of range measurements. A variety of methods have been used to compute these residual errors. An approximate estimate of range accuracy is obtained from the following:

$$\text{Estimated range error} = \sigma = \sqrt{\frac{\sum_{i=1}^{n} v_i^2}{n - 1}} \tag{8.3}$$

where n is the number of observed ranges and $\sum_{i=1}^{n} v_i^2$ is the sum of the squared residuals.

Adding redundant ranges will not necessarily make a significant improvement in the positional accuracy because inherent random and systematic errors are still present. It will, however, help detect the existence of large systematic errors that might have otherwise gone undetected using a nonredundant range–range system.

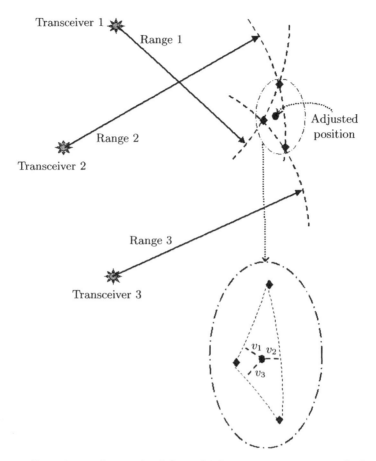

Fig. 8.2 Fundamentals of the multiple-range intersection method.

8.2 GPS TECHNIQUES

The global positioning system or GPS (Figure 8.3) has rapidly become the standard surveying and navigation mode, replacing microwave and other types of ranging systems. Real-time GPS positional accuracies now exceed those of any other positioning system. Most significantly, GPS does not require the time-consuming calibration that microwave equipment does. Numerous public and private differential GPS systems now exist which give a broad coverage with an accuracy in excess of standard nondifferential GPSs.

Actually, GPS is a real-time, all-weather, 24-hour, worldwide, three-dimensional absolute satellite-based positioning system. This system consists of two positioning services:

- The precise positioning service (PPS) was developed for the military and other authorized users and provides an accuracy of 5–10 m in absolute positioning mode.

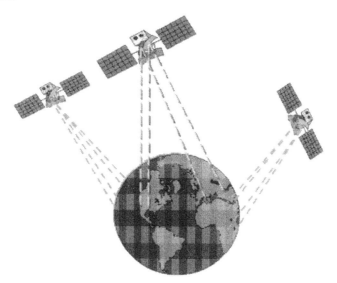

Fig. 8.3 GPS with satellite navigation.

- The standard positioning service (SPS) is available to civilian users and provides an accuracy of 10–20 m in absolute positioning mode.

8.2.1 Differential GPS (DGPS)

For many applications, absolute positioning does not provide sufficient accuracy. Differential GPS is a technique which can provide relative positioning with an accuracy from a few meters to a few millimeters, depending on the DGPS method used. Such a scheme utilizes an additional reference signal, transmitted from a known location, which is used to reduce the inherent error of standard GPS. Normally, DGPS utilizing code phase measurements can provide a relative accuracy of a few meters. However, DGPS utilizing carrier-phase measurements can provide a relative accuracy of a few centimeters. Differential GPS requires two or more GPS receivers to be recording measurements simultaneously. With two stations recording observations at the same time, GPS processing software can reduce or eliminate common-mode errors. Both code and carrier-phase DGPS can be performed in real time, making it applicable for moving platforms.

8.2.2 GPS tracking modes

There are basically two general modes which are used to determine the distance between a GPS satellite and a ground-based receiver antenna. These measurements are made by signal-phase comparison techniques:

- carrier-phase tracking;
- code-phase tracking.

Either the satellite's carrier frequency phase or the phase of a digital code modulated on the carrier phase may be used, or tracked, to resolve the distance between the satellite and the receiver. The resulting positional accuracy is dependent upon the tracking method used.

GPS satellites actually broadcast on two carrier frequencies: L1 at 1575.42 MHz (19 cm wavelength) and L2 at 1227.60 MHz (24 cm wavelength). Modulated on these frequencies are the coarse acquisition (C/A) (300 m wavelength) and the precise (P) (30 m wavelength) codes. In addition, a 50 bps satellite navigation message containing the ephemeris and health status of each satellite is transmitted. The C/A and P codes are both present on the L1 frequency. Only the P code is present on the L2 frequency.

The higher frequencies of the carrier signal (L-band) have wavelengths of 19 and 24 cm, from which a distance can be resolved through post-processing software to approximately 2 mm. The modulating code has a wavelength of 300 m and will only yield distances accurate to about 1 m.

8.2.3   GPS error sources

The accuracy of GPS is a function of the error and interference on the GPS signal and the processing technique used to reduce and remove these errors. The same types of phenomena as found in range–range microwave systems affect GPS signals. Both types of systems are highly affected by humidity and multipath.

8.2.3.1   Tropospheric error   Humidity is included in this type of error. Humidity can delay a time signal up to approximately 3 m. Satellites low on the horizon will be sending signals across the face of the earth through the troposphere. Satellites directly overhead will transmit through much less troposphere. Masking the horizon angle to 15° can minimize tropospheric error.

8.2.3.2   Ionospheric error   Sunspots and other electromagnetic phenomena cause errors in GPS range measurements of up to 30 m during the day and as high as 6 m at night. The errors are not predictable, but can be estimated.

8.2.3.3   Multipath   Multipath is the reception of reflected, refracted, or diffracted signals in lieu of a direct signal. Multipath signals can occur below or above the antenna. Multipath magnitude is less over water than over land, but it is still present and always changing. If possible the placement of the GPS receiver antenna should avoid areas where multipath is more likely to occur (e.g., rock outcrops, metal roofs, commercial roof-mounted heating and air conditioning, buildings, cars, ships, etc.). Increasing the height of the antenna is one method of reducing multipath at a reference station. Multipath occurrence on a satellite transmission can last several minutes while the satellite passes overhead. Masking out satellite signals from the horizon up to 15° will also reduce multipath effects.

8.3 POSITIONING TECHNIQUES

In this section we will consider positioning techniques that are based on a different classification.

8.3.1 Introduction

According to the place where measurements and their evaluation take place, positioning systems can be classified as network-based, handset-based, or hybrid [87–89].

8.3.1.1 Network-based systems In network-based systems, calculation of the position is computed in a control station which is located in a remote position from the object to be positioned. More specifically, a set of stationary receivers make appropriate measurements of a signal originating from or reflecting off the object to be positioned. These measurements are then used in the control station where the object's location is computed. Such systems involve lower mobile terminal costs than the self-positioning technique, but they are usually less accurate than handset-based systems.

8.3.1.2 Handset-based systems Handset-based systems involve the calculation of the position of the object to be positioned itself. More specifically, the positioning receiver uses appropriate measurements made from signals sent by location-known transmitters in order to determine its own position. The main advantage of handset-based systems is that this technology can better address privacy issues as the location is computed and stored in the receiver. However, mobile terminal costs will increase as additional signal processing is integrated in the receiver.

8.3.1.3 Hybrid systems Finally, hybrid systems incorporate a combination of handset and network-based technologies. Usually, the object to be positioned takes the measurements and transmits the results to the stationary network where the object's location is estimated. The main aim of this method is to produce a more robust estimate of location in a single process.

8.3.2 Network-based techniques

8.3.2.1 Received signal strength (RSS) With this technique the signal strength of the object to be positioned is measured at several stationary receivers. Ideally, each measurement will provide a circle of radius representing the distance between the object and the receiver that made this measurement, centered at the corresponding receiver. The object position is then given by the intersection of these circles. In two-dimensional positioning and assuming that no measurement error occurs, at least three circles are required in order to resolve the ambiguities arising from multiple crossings of the circles.

The accuracy of the position can be improved by increasing the number of measurements and then averaging the results. However, this approach requires an exact knowledge of the path loss in order to get an accurate estimation of the signal strength at the receivers. In a multipath fading environment, it is difficult to relate the distance with the received signal strength. Figure 8.4 shows an example of RSS positioning. Obviously, it is rare that all three distorted circles coincide exactly, and ambiguity in the coincidence point determines the systematic positioning error.

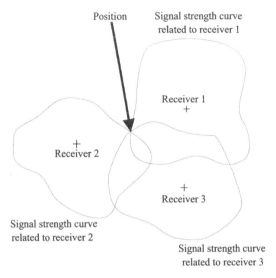

Fig. 8.4 In RSS positioning the intersections of the distorted circles determine the position of the object.

8.3.2.2 Angle of arrival (AOA)/Direction of arrival (DOA) In this method the angle of arrival of the signal sent by the object to be positioned is measured at several stationary receivers by steering the main lobe of a directional antenna or an adaptive antenna array. Each measurement forms a radial line from the receiver to the object to be positioned. In two-dimensional positioning the position of the object is defined at the intersection of two directional lines of bearing. In practice, more than two receivers may be employed to combat inaccuracies introduced by multipath propagation effects.

This method has the advantage of not requiring synchronization of the receivers nor an accurate timing reference. On the other hand, receivers require regular calibration in order to compensate for temperature variations and antenna mismatches. This can be done automatically.

Assuming that we know the coordinates of two receivers the derivation of the position is straightforward. Without loss of generality, we can assume that these coordinates are $(0,0)$ for receiver 1 and $(0, y_2)$ for receiver 2. Given α and β, respectively, as the angles of arrival of the signal from the object at receiver 1 and receiver 2 we can define two straight lines by

$$y = \tan(\alpha)x \tag{8.4}$$
$$y = \tan(\beta)x + y_2 \tag{8.5}$$

Substituting Equation (8.4) into Equation (8.5) yields

$$x = x_0 = \frac{y_2}{\tan(\alpha) - \tan(\beta)} \tag{8.6}$$

Substituting x_0 into Equation (8.4) gives us a unique y_0, and thus the point defined by the coordinates (x_0, y_0) is the desired position. Figure 8.5 demonstrates an example of AOA positioning.

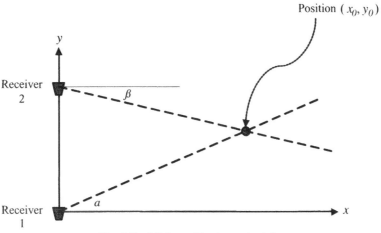

Fig. 8.5 AOA positioning principle.

8.3.2.3 *Time of arrival (TOA)* With this approach the time that the signal sent by the object to be positioned is measured at each receiver. Given that the propagation time of the signal is known and is directly proportional to its traversed distance, the measured time can provide a circle of radius representing the distance between the object and the receiver, centered at the latter. In two-dimensional positioning, at least three circles are required. This technique is easy to implement; however, it requires knowledge of the transmission time of the emitted signal as well as synchronization of the transmitter and receivers clocks. Otherwise, huge position errors can occur. For example, a clock inaccuracy of just 1 μs will lead to a position error of 300 m. Furthermore, this technique can suffer from multipath propagation effects.

Assuming that we know the coordinates of three receivers, the derivation of the position is simple. Without loss of generality, we can assume that these coordinates are as follows:

$$\text{Receiver 1} : (0, 0)$$
$$\text{Receiver 2} : (0, y_2)$$
$$\text{Receiver 3} : (x_3, y_3)$$

Given that t_1, t_2, and t_3 denote the time it takes the signal to travel from the object to be positioned to the respective receivers and that c denotes the signal's speed of propagation, the distances between the object and each of the receivers are given by

$$d_1 = c.t_1 = \sqrt{x^2 + y^2} \tag{8.7}$$

$$d_2 = c.t_2 = \sqrt{x^2 + (y - y_2)^2} \tag{8.8}$$

$$d_3 = c.t_3 = \sqrt{(x - x_3)^2 + (y - y_3)^2} \tag{8.9}$$

Each of these equations defines a circle whose x and y are unknowns. Squaring both sides of the above equations and some basic manipulation yields

$$y = \frac{y_2^2 + d_1^2 - d_2^2}{2y_2} \tag{8.10}$$

Substituting Equation (8.10) into Equation (8.7) gives two values for x. Only one of these values which is positive is correct, and this way we have calculated the desired position. Figure 8.6 demonstrates an example of TOA positioning.

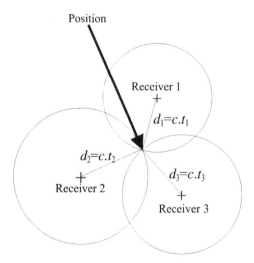

Fig. 8.6 TOA positioning principle.

8.3.2.4 Time difference of arrival (TDOA) In this technique [90] the difference in time at which the signal from the object to be positioned arrives at two different receivers is measured. Each time difference is then converted into a hyperboloid with a constant distance difference between the two receivers. In two-dimensional positioning, at least two pairs of receivers are required; the position is the intersection of the two corresponding hyperboloids.

This technique requires synchronization of the receivers' clocks; however, knowledge of the absolute transmission time of the emitted signal is no longer required. Here again, multipath propagation effects can influence the accuracy of the position and the location of the receivers. It is important to note that the wider the bandwidth of the signal, the lower the measurement error becomes.

The derivation of the position is straightforward if we assume that we know the coordinates of three receivers. Without loss of generality, it can be assumed again that these coordinates are

$$\text{Receiver } 1 : (0,0)$$
$$\text{Receiver } 2 : (0, y_2)$$
$$\text{Receiver } 3 : (x_3, y_3)$$

Given that t_1, t_2, and t_3 denote the time it takes the signal to travel from the object to be positioned to the corresponding receivers, the distances between the object and each of the receivers are given by

$$d_1 = c.t_1 \tag{8.11}$$
$$d_2 = c.t_2 \tag{8.12}$$
$$d_3 = c.t_3 \tag{8.13}$$

We can now define two hyperboloids using the TDOA algorithm, that is,

$$d_{1,2} = d_2 - d_1 = c.(t_2 - t_1)$$
$$= \sqrt{x^2 + (y - y_2)^2} - \sqrt{x^2 + y^2} \tag{8.14}$$

and

$$d_{1,3} = d_3 - d_1 = c.(t_3 - t_1)$$
$$= \sqrt{(x - x_3)^2 + (y - y_3)^2} - \sqrt{x^2 + y^2} \tag{8.15}$$

where x and y are unknowns. Taking the square in Equations (8.14) and (8.15) yields

$$2d_{1,2}\sqrt{x^2 + y^2} = y_2^2 - d_{1,2}^2 - (2y_2)y \tag{8.16}$$
$$2d_{1,3}\sqrt{x^2 + y^2} = x_3^2 + y_3^2 - d_{1,3}^2 - (2x_3)x - (2y_3)y \tag{8.17}$$

Knowing that $x^2 + y^2$ is not equal to zero, Equations (8.16) and (8.17) lead to

$$x = by + a \tag{8.18}$$

where

$$b = \frac{2y_2 d_{1,3} - 2y_3 d_{1,2}}{2x_3 d_{1,2}} \tag{8.19}$$

and

$$a = \frac{x_3^2 d_{1,2} + y_3^2 d_{1,2} - y_2^2 d_{1,3} + d_{1,2}^2 d_{1,3} - d_{1,2} d_{1,3}^2}{2x_3 d_{1,2}} \tag{8.20}$$

Substituting Equation (8.18) into Equation (8.16) gives

$$2d_{1,2}\sqrt{(b^2 + 1)y^2 + (2ba)y + a^2} = y_2^2 - d_{1,2}^2 - (2y_2)y \tag{8.21}$$

which results in

$$\left[4d_{1,2}^2(b^2 + 1) - 4y_2^2\right]y^2 + \left[8bad_{1,2}^2 + 4(y_2^2 - d_{1,2}^2)y_2\right]y$$
$$+ \left[4a^2 d_{1,2}^2 - (y_2^2 - d_{1,2}^2)^2\right] = 0 \tag{8.22}$$

Equation (8.22) is a quadratic equation with two roots that are the y-coordinates of the intersection points of the hyperboloids. Using Equation (8.18) provides the corresponding x-coordinates. To remove the ambiguity, we can define another hyperboloid by

$$d_{2,3} = d_3 - d_2 = c.(t_3 - t_2)$$
$$= \sqrt{(x - x_3)^2 + (y - y_3)^2} - \sqrt{x^2 + (y - y_2)^2} \tag{8.23}$$

Substitution of $d_{1,3}$ by $d_{2,3}$ in Equations (8.16)–(8.22) yields two points. One of these points matches the previous ones. This point is the required position. Figure 8.7 demonstrates an example of TDOA positioning.

8.3.2.5 Multipath fingerprinting This positioning technique is based on matching the received signal fingerprint to a reference fingerprint previously measured for a known location in the network and stored in a central database. Each spot in a network would have a unique signature in terms of TOA, AOA, and RSS, observed from at least one receiver. Therefore, to build the database we have to divide the service area into nonoverlapping zones and record the received signal pattern corresponding to each zone in the database. This technique performs better than other techniques in multipath-rich environments, but the size of the database increases considerably when the service area becomes large.

8.3.3 Handset-based techniques

These are satellite-based positioning techniques, such as GPS, Galileo, or the global navigation satellite system (GLONASS). Currently, there are three potential global satellite navigation systems:

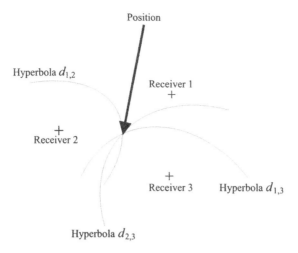

Fig. 8.7 TDOA positioning principle.

- the US-made GPS which is relatively accurate and reliable (described in Section 7.2);

- the European Galileo is still in development, but promises to be more accurate than the GPS system;

- the Russian GLONASS.

A receiver using such systems requires a clear view of the sky and signals from at least four satellites that are part of a constellation. For three-dimensional positioning, three time measurements (just like the TOA method) are used to define a sphere in space centered at the corresponding satellite and a fourth measurement is required to solve receiver clock bias caused by the unsynchronized clocks of the satellite and the receiver. Furthermore, each satellite sends its almanac to the GPS receiver to describe its position in the sky.

Such systems are most unreliable in buildings or urban environments, as the receiver might not be able to get four satellites in its LOS or receive any satellite at all. Furthermore, without knowledge of the state of the satellite constellation a receiver may take several minutes to obtain a measured position.

As described before, differential GPS (DGPS) is used to improve the accuracy of a conventional GPS receiver. Along with the mobile receiver a stationary receiver is used to receive signals from all satellites that are visible from the station. A position measurement is made for each received signal, and correction data for each satellite are processed as the station's location is known. The data are then sent to the mobile GPS receiver which can improve the accuracy of the measured position by picking up information from those satellites it uses for the measurement. This method corrects errors due to atmosphere or clock bias, but cannot remove errors due to multipath-rich environments as the station cannot predict the mobile's nearby surroundings.

The usual way to calculate correction data is to measure the pseudo-range error of each satellite. This technique is based on the principle that the received signal from a satellite at two different receivers on earth located a few kilometers apart will suffer almost the same errors and delays due to the atmosphere. As the exact position of the stationary receiver is known as well as the true distance to the satellite from the stored almanac, it is possible to theoretically calculate how long it takes for the signal to reach the receiver. Compared with the measured time, the difference gives an error correction time factor which can be used to correct the pseudo-range given by

$$\text{Corrected pseudorange} = c \cdot (t_{\text{measured}} - t_{\text{correction}}) \qquad (8.24)$$

where t_{measured} denotes the measured time for the signal to arrive and $t_{\text{correction}}$ represents the time correction factor. This corrected pseudo-range is then sent to the mobile GPS receiver which can then correct its own measurements.

The final technique discussed here uses the enhanced observed time difference (E-OTD) technique, which is similar to the TDOA method. The mobile object to be positioned listens to bursts sent from neighboring stationary transmitters with accurate timing sources and measures the time difference of arrival from these transmitters. The object to be positioned then uses these measurements to determine its own location through a trilateration scheme, as in the TDOA method. This technique requires at least two pairs of synchronized transmitters in order to remove the ambiguity in position.

8.3.4 Hybrid techniques

8.3.4.1 Assisted GPS (A-GPS)   In this technique a reference GPS receiver, which sees the same satellites as the object to be positioned, is used. With this reference GPS receiver the network can predict the GPS signal that the object will receive and, thus, assist the object to be positioned by sending information about the position of the GPS satellites. The object can then make quick GPS measurements (the acquisition delays of a conventional GPS are eliminated) and send them back to the network where the exact position is estimated. Furthermore, the performance of conventional GPS receivers in low SNR conditions is improved with A-GPS technology.

8.3.4.2 Advanced forward link trilateration (AFLT)   This technique is similar to the TDOA method. The object to be positioned measures the time of arrival of signals from the transmitters and reports them back to the network where the calculation of the object's position is processed. Here, at least three transmitters are required to get an optimal position fix. The accuracy of this method is usually better than using the TDOA technique as all the computations are made in the network.

8.3.5 Other techniques

As each technique has its own advantages and disadvantages, it is possible to combine several of these techniques in order to improve the accuracy or the cost of the positioning process. Some examples of combination are AOA+RSS, AOA+TDOA, E-OTD+A-GPS, and A-GPS+AFLT.

8.4 TIME RESOLUTION ISSUES

As time-based techniques are frequently used in geolocation, one of the main sources of error is due to the time resolution of the signal used to make the appropriate measurements. These signals can be classified into four main systems: narrowband, wideband, super-resolution, or UWB ranging.

8.4.1 Narrowband systems

Usually, in the narrowband ranging technique the time-based approach is used and the distance between two points is determined by measuring the phase difference between transmitted and received signals. The relation between the phase of a received signal ϕ and the time of arrival of this signal t is given by

$$t = \frac{\phi}{w_0} \tag{8.25}$$

where w_0 denotes the frequency of the signal in radians per second. Narrowband ranging techniques can reach accuracies of the order of 1 m, like the DGPS system, but a direct LOS path is required in order to obtain accurate results. In multipath-rich environments, such as urban or indoor propagation scenarios where a direct LOS path is often not available, substantial measurement errors occur. In fact, the received signal is the sum of all the signals arriving from different paths with different amplitudes and delays. Thus, if a direct LOS path is not present, then the received signal arriving from a NLOS scenario takes more time to reach the receiver than expected in a LOS scenario. The delay thus observed is the source of error in narrowband ranging techniques.

8.4.2 Wideband systems

Another widely used technique is the wideband signal approach where the DSSS method is the most commonly used form, as this technique performs better than competing systems at suppressing interference. In such a system a known PN signal, which is modulated using a modulation technique (such as BPSK, QPSK, etc.), is multiplied by the carrier signal, which is thus replaced by a wide bandwidth signal with a spectrum equivalent to that of the noise signal.

Without loss of information, this technique allows the signal power to be less than or equal to the noise power. Usually, in order to measure the time of arrival of the signal a sliding correlator or a matched filter is used at the receiver which cross-correlates the received signal with a stored reference PN sequence. The arrival time of the first correlation peak is used as the time measurement.

In the time-based approach the resolution of the estimation is related to the bandwidth of the spread signal by

$$d = \frac{c}{\text{BW}} \tag{8.26}$$

where d denotes the absolute resolution and BW is the bandwidth of the spread signal. For example, a signal of 300 MHz bandwidth allows an absolute resolution of 1 m to be obtained. This is true when a direct LOS path is visible. However, when multipath propagation occurs, wideband ranging techniques cannot maintain this range of accuracy; even so, it is still better than the narrowband technique.

8.4.3 Super-resolution techniques

This method has been introduced in order to get higher ranging accuracy for a given bandwidth. The time of arrival of a signal can be determined with high resolution using a frequency domain super-resolution technique [91]. This technique uses the frequency response of the channel.

The impulse response of the channel can be modeled as

$$h(t) = \sum_{k=0}^{L_p - 1} \alpha_k \delta(t - \tau_k) \tag{8.27}$$

where L_p denotes the total number of multipath components and α_k and τ_k are the complex attenuation and propagation delay of path number k, respectively. Furthermore

$$\alpha_k = |\alpha_k| e^{j\theta_k} \tag{8.28}$$

where θ_k represents the phase of complex attenuation. In the model, as propagation delays τ_k, $0 \le k \le L_p - 1$, are in ascending order, the parameter τ_0 represents the propagation delay of the shortest path, which is the path of the direct LOS. This delay has to be estimated as it is required for computing the time of arrival. Then, taking the Fourier transform of Equation (8.27), the frequency domain channel response can be obtained as

$$H(f) = \sum_{k=0}^{L_p - 1} \alpha_k e^{-j2\pi f \tau_k} \tag{8.29}$$

In the model, parameters α_k and τ_k are time-invariant random variables since the motion of the obstacles (people, objects, etc.) is very slow compared with the measurement time interval. The phase of complex attenuation θ_k is assumed to be a random variable uniformly distributed in $[0, 2\pi]$.

In practice, the frequency domain channel response can be obtained by deconvolution of the received signal of a DSSS system over the frequency band of high SNR, by sweeping the channel at different frequencies or by using a multicarrier modulation technique. TOA estimation is then performed by first interchanging the role of time and frequency in Equation (8.29). The result is the harmonic signal model as follows:

$$H(\tau) = \sum_{k=0}^{L_p-1} \alpha_k e^{-j2\pi f_k \tau} \tag{8.30}$$

Actually, a spectral estimation technique, such as the multiple signal classification (MUSIC) algorithm, can compute a time domain analysis from the frequency response of the channel. Hence, by sampling the channel frequency response $H(f)$ at L equally spaced frequencies we can get some measurement data. Taking into account the AWGN $w(l)$ with zero mean and variance σ_w^2, the sampled frequency response of the channel can be rewritten as

$$\begin{aligned} x(l) &= \hat{H}(f_l) \\ &= H(f_l) + w(l) \\ &= \sum_{k=0}^{L_p-1} \alpha_k e^{-j2\pi(f_0+l\Delta f)\tau_k} + w(l) \end{aligned} \tag{8.31}$$

where $l = 0, 1, \ldots, L-1$. Let us now rewrite Equation (8.31) in vector form as

$$\begin{aligned} \mathbf{x} &= \mathbf{H} + \mathbf{w} \\ &= \mathbf{V}\boldsymbol{\alpha} + \mathbf{w} \end{aligned} \tag{8.32}$$

where

$$\mathbf{x} = [x(0)\ x(1)\ \ldots\ x(L-1)]^{\mathrm{T}} \tag{8.33}$$
$$\mathbf{H} = [H(0)\ H(1)\ \ldots\ H(L-1)]^{\mathrm{T}} \tag{8.34}$$
$$\mathbf{w} = [w(0)\ w(1)\ \ldots\ w(L-1)]^{\mathrm{T}} \tag{8.35}$$
$$\mathbf{V} = [\mathbf{v}(\tau_0)\ \mathbf{v}(\tau_1)\ \ldots\ \mathbf{v}(\tau_{L_p-1})] \tag{8.36}$$
$$\boldsymbol{\alpha} = [\alpha_0'\ \alpha_1'\ \ldots\ \alpha_{L_p-1}']^{\mathrm{T}} \tag{8.37}$$

and

$$\mathbf{v}(\tau_k) = [1\ e^{-j2\pi\Delta f\tau_k}\ \ldots\ e^{-j2\pi(L-1)\Delta f\tau_k}]^{\mathrm{T}} \tag{8.38}$$
$$\alpha_k' = \alpha_k e^{-j2\pi f_0 \tau_k} \tag{8.39}$$

where the superscript T denotes the matrix transpose operation. The MUSIC algorithm detects frequencies by performing an eigen-decomposition on the covariance matrix of the data vector described in Equation (8.32):

$$\begin{aligned} \mathbf{R}_{xx} &= \mathrm{E}\{\mathbf{x}\mathbf{x}^{\mathrm{H}}\} \\ &= \mathbf{V}\mathbf{A}\mathbf{V}^{\mathrm{H}} + \sigma_w^2 \mathbf{I} \end{aligned} \tag{8.40}$$
$$\tag{8.41}$$

where $\mathrm{E}\{\cdot\}$ is the expectation,

$$\mathbf{A} = \mathrm{E}\{\boldsymbol{\alpha}\boldsymbol{\alpha}^{\mathrm{H}}\} \tag{8.42}$$

$\mathbf{I}$ is the identity matrix, and superscript H denotes the conjugate transpose operation. Since propagation delays τ_k in Equation (8.27) are all different, the column vectors of $\mathbf{V}$ are linearly independent.

Given that the magnitude of α_k is constant and the phase θ_k is a uniform random variable in $[0, 2\pi]$, the $L_p \times L_p$ covariance matrix $\mathbf{A}$ is nonsingular. Hence, assuming $L > L_p$ the rank of the matrix $\mathbf{VAV}^{\mathrm{H}}$ is L_p. This means that the $L - L_p$ smallest eigenvalues of $\mathbf{R}_{xx}$ are all equal to σ_w^2 and are thus called noise eigenvectors, while the L_p largest eigenvalues are called signal eigenvectors. The signal vector $\mathbf{x}$ is contained in an L-dimensional subspace that can therefore be split into a signal subspace and a noise subspace. These subspaces are orthogonal and are defined by signal eigenvectors and noise eigenvectors, respectively.

Assuming that the eigenvectors are normalized, we can write

$$\mathbf{Q}_w^{\mathrm{H}}\mathbf{Q}_w = \mathbf{I} \tag{8.43}$$

where

$$\mathbf{Q}_w = [\mathbf{q}_{L_p}\ \mathbf{q}_{L_p+1}\ \cdots\ \mathbf{q}_{L-1}] \tag{8.44}$$

and $\mathbf{q}_k$, $L_p \leq k \leq L-1$, are the noise eigenvectors. The projection matrix of the noise subspace is then given by

$$\mathbf{P}_w = \mathbf{Q}_w(\mathbf{Q}_w^{\mathrm{H}}\mathbf{Q}_w)^{-1}\mathbf{Q}_w^{\mathrm{H}} \tag{8.45}$$

$$= \mathbf{Q}_w\mathbf{Q}_w^{\mathrm{H}} \tag{8.46}$$

Since the signal subspace is orthogonal to the noise subspace and the vector $\mathbf{v}_{\tau_k}$, $0 \leq k \leq L_p - 1$, belongs to the signal subspace we have

$$\mathbf{P}_w\mathbf{v}_{\tau_k} = 0 \tag{8.47}$$

The above equation means that the vector $\mathbf{v}_{\tau_k}$, $0 \leq k \leq L_p - 1$, is orthogonal to the noise subspace. Then, given that the projection matrix is idempotent, that is,

$$\mathbf{P}_w^{\mathrm{H}}\mathbf{P}_w = \mathbf{Q}_w\mathbf{Q}_w^{\mathrm{H}}\mathbf{Q}_w\mathbf{Q}_w^{\mathrm{H}} = \mathbf{Q}_w\mathbf{I}\mathbf{Q}_w^{\mathrm{H}} \tag{8.48}$$

$$= \mathbf{Q}_w\mathbf{Q}_w^{\mathrm{H}} \tag{8.49}$$

$$= \mathbf{P}_w \tag{8.50}$$

the multipath delays τ_k, $0 \leq k \leq L_p - 1$, correspond to the delays when the time domain MUSIC pseudo-spectrum, defined as

$$\begin{aligned}
S_{\mathrm{MUSIC}}(\tau) &= \frac{1}{\|\mathbf{P}_w\mathbf{v}(\tau)\|^2} = \frac{1}{\mathbf{v}(\tau)^{\mathrm{H}}\mathbf{P}_w^{\mathrm{H}}\mathbf{P}_w\mathbf{v}(\tau)} \\
&= \frac{1}{\mathbf{v}(\tau)^{\mathrm{H}}\mathbf{P}_w\mathbf{v}(\tau)} = \frac{1}{\|\mathbf{Q}_w^{\mathrm{H}}\mathbf{v}(\tau)\|^2} \\
&= \frac{1}{\sum_{k=L_p}^{L-1} |\mathbf{q}_k^{\mathrm{H}}\mathbf{v}(\tau)|^2}
\end{aligned} \tag{8.51}$$

is maximum.

In practice, the channel frequency response is estimated using the received signal, and the super-resolution algorithm then performs the time domain pseudo-spectrum transformation as in Equation (8.51). The time of arrival is finally estimated by detecting the first peak of the obtained pseudo-spectrum. However, in a high multipath environment, where NLOS conditions occur between the transmitter and the receiver, this technique cannot eliminate ranging errors at some locations.

It should be noted that the MUSIC algorithm is only an example and other techniques can be applied to the super-resolution algorithm.

8.4.4 UWB systems

Finally, the most recent, accurate, and promising technique is the UWB approach. We should note that ranging accuracy depends upon signal bandwidth. The larger the bandwidth, the better the accuracy of the estimation of the time of arrival. In fact, as the bandwidth of UWB systems is usually in excess of 2–3 GHz the ranging accuracy is of the order of 1 cm. This fact is clear from Equation (8.26).

The large bandwidth of UWB systems means that they are able to resolve multiple paths and combat multipath fading and interference. However, such systems have a limited range and building penetration, due to the high attenuation associated with the high-frequency content of the signal. It was shown by Fontana and Gunderson in [92] and [93] that a set of fixed UWB beacons can provide a sub-centimeter ranging accuracy in a multipath scenario.

Another method uses a sub-sampled version of the received signal as in [94]. A realistic UWB channel can be modeled as

$$h(t) = \sum_{l=0}^{L-1} a_l p_l(t - t_l) \qquad (8.52)$$

where t_l denotes a signal delay along the lth path, a_l is a complex propagation coefficient corresponding to this path, and $p_l(t)$ are different pulse shapes that correspond to different paths. Hence, if a signal $s(t)$ is transmitted over this channel the spectral coefficients of the received signal $y(t)$ are given by

$$Y[n] = S[n] \sum_{l=0}^{L-1} P_l[n] a_l e^{-j\omega_n t_l} + N[n] \qquad (8.53)$$

where $N[n]$ represents the spectral coefficients of the noise and $P_l[n]$ are now the unknown parameters. These parameters can be approximated with polynomials of degree $D \leq R - 1$, that is,

$$P_l[n] = \sum_{r=0}^{R-1} p_{l,r} n^r \qquad (8.54)$$

Therefore, Equation (8.53) can be rewritten as

$$Y[n] = S[n] \sum_{l=0}^{L-1} a_l \sum_{r=0}^{R-1} p_{l,r} n^r e^{-j\omega_n t_l} + N[n] \tag{8.55}$$

Then, if we define

$$c_{l,r} = a_l p_{l,r} \tag{8.56}$$

and

$$Y_s[n] = \frac{Y[n]}{S[n]} \tag{8.57}$$

we get

$$Y_s[n] = \sum_{l=0}^{L-1} \sum_{r=0}^{R-1} c_{l,r} n^r e^{-j\omega_n t_l} + N[n] \tag{8.58}$$

Now, an annihilating filter for $Y_s[n]$ is

$$H(z) = \prod_{l=0}^{L-1} (1 - e^{-j\omega_0 t_l} z^{-1})^R$$

$$= \sum_{k=0}^{RL} H[k] z^{-k} \tag{8.59}$$

with multiple roots at

$$z_l = e^{-j\omega_0 t_l} \tag{8.60}$$

where ω_0 denotes the sampling frequency. Then, noting that the received signal $y(t)$ can be modeled as a convolution of L impulses with a known data sequence $g(t)$ we obtain

$$y(t) = \sum_{l=0}^{L-1} a_l p_l(t - t_l) * g(t) \tag{8.61}$$

and

$$Y[n] = \sum_{l=0}^{L-1} a_l P_l[n] G[n] e^{-jn\omega_c t_l} \tag{8.62}$$

where

$$\omega_c = \frac{2\pi}{T_c} \tag{8.63}$$

In the above equation T_c denotes the cycle time. If we use the polynomial approximation of coefficients $P_l[n]$ as defined in Equation (8.54) the minimum sampling rate corresponds to the total number of degrees of freedom per cycle (i.e. $2RL$). Therefore, accurate delay estimation can be performed by increasing the sampling rate over the entire cycle.

8.5 UWB POSITIONING AND COMMUNICATIONS

Due to the fine time resolution associated with UWB signals and, therefore, the possibility to have simultaneous timing, location, ranging, and communications integrated in a single UWB system, it is an attractive candidate for combined future wireless communication and location positioning systems [95,96]. Indeed, by transferring information for both location and control, such systems could extend the senses of people or machines into their own environment.

8.5.1 Potential user scenarios

8.5.1.1 Intelligent wireless area network (IWAN)   As the current generation of narrowband networks does not enable context-aware services, such as asset tracking, alarm zones, etc., UWB positioning devices in a master–slave topology could be used in an IWAN to enable such context-aware services. A high density of devices (at least five per room) communicating at a medium to low data rate combined with positioning capability is used in an IWAN, which can cover medium to long distances. To be reliable in a smart home or office, such devices have to be very low cost and have very low power consumption as they have to be integrated into all the objects and assets that need to be intelligently controlled. Furthermore, a wireless bridge with the outside world could be implemented providing the capability to remotely control the sensors.

8.5.1.2 Sensor, positioning, and identification network (SPIN)   This scenario is more suitable for industrial factories or warehouses because of the very specific interference and propagation environment attributed to these buildings. As this environment continuously changes, adaptive systems and numerous links of reliability are required. Therefore, a SPIN uses a very high density of devices (hundreds per floor) communicating at a low data rate combined with positioning capability. These devices cover medium to long distances to a master station in the usual master–slave topology. However, an ad-hoc topology is also possible.

8.5.2 Potential applications

Nowadays, as wireless communications depart from a centralized server system, the position location technology associated with communication devices is becoming a requirement for many applications.

8.5.2.1 Personal location   Positioning devices could be used in personal location systems when the user needs to precisely locate someone or something. For example, police officers or firefighters in action could know the position of colleagues or drivers could easily locate their cars in a large car park. Many other personal location applications exist where precise positioning would be required.

8.5.2.2 Inventory control   In such applications, positioning devices could act as bar code identification tags and could give the precise location of the inventory item at the same time. Therefore, as an object is being moved out of stock, inventory control is instantly notified and, thus, real-time information on the stock is possible.

8.5.2.3 Machine control   Another application for positioning devices is to combine them with small robotic vehicles which could be used for difficult access infrastructure inspection, such as bridges or sewers, or in a harmful environment, such as nuclear plants. Indeed, due to its accurate location capability a robot could be monitored at a safe, remote location. Another possibility is to use reliable small robots in a house or office environment in order to perform some tasks without the need for supervision.

8.5.2.4 Smart highways   Positioning devices could be used to assist autopilots in automobiles. Indeed, vehicles could be guided along a highway by integrated UWB sensors along the road. Furthermore, such sensors placed inside the vehicles might enable them to communicate and, thus, provide real-time local intelligence in order to avoid accidents.

8.5.2.5 Smart homes and offices   Integrating UWB sensors into home or office appliances, such as televisions, lamps, computers, etc., might enable technologies that turn the desired appliances on or off by knowing the location of people in the home or the office as well as the location of the appliances.

8.5.2.6 Ad-hoc networking   An ad-hoc network has no infrastructure and consists of several mobile terminals that can communicate with each other without fixed routers, even at the base stations. Each mobile node can act as a terminal and as a router. So, it can find and maintain a suitable route to other nodes in the network dynamically. Such a feature provides a remarkable increase in the level of autonomy compared with the traditional fixed communication infrastructure. However, the location of the mobile, the strict constraints for power consumption of battery-powered terminals, and multipath interferences are the main concerns in ad-hoc networks.

When using UWB radio technology, two mobile terminals inside the network can determine their distance within 10 cm and arrival time delays at the receiver as small as a fraction of a nanosecond. Ad-hoc networking using UWB technology is a novel application that is able to overcome the main limitations of traditional multihop solutions, such as power constraints, multipath propagation, and location of the mobile terminal.

8.6   SUMMARY

In this chapter the fundamentals of UWB positioning and location were described. The application of various wireless positioning techniques in terms of the operating frequency of the system was explained. Practical considerations, such as antenna and multipath effects, were briefly considered. The standard surveying and navigation systems, GPS and differential GPS, and their major error sources were addressed.

We also classified positioning systems according to the place where measurements and their evaluation take place, such as network-based, handset-based, or hybrid. Time resolution issues for narrowband and wideband systems as well as super-resolution techniques were discussed.

Finally, the most recent and accurate technique, the UWB approach, was explained. The combination of UWB positioning and communication as well as some potential applications concluded the chapter.

Problems

Problem 1. What are the effects of the following parameters on the performance of a positioning system?

(i) frequency;

(ii) bandwidth; and

(iii) multipath.

Problem 2. How are the following network-based techniques of location finding compared?

(i) received signal strength;

(ii) angle of arrival;

(iii) time of arrival;

(iv) time difference of arrival; or

(v) multipath fingerprinting.

Problem 3. Can you suggest some other possible applications of personal location using UWB systems?

Problem 4. How do UWB positioning systems differ from conventional location systems?

9

Applications using UWB systems

There have been many UWB applications in both the military and government area that have already been developed and shown to be viable products. Although this chapter will only provide an overview of a select few applications, it should give the reader an idea of what has already been accomplished and what applications are planned for the future in the commercial sector. In particular, we want to point out that, following the plan of this book, we do not examine radar applications, which are perhaps the most developed of all UWB applications. Rather, we wish to focus on communications, looking ahead to their use in consumer products.

As test cases for both military and commercial use we examine the precision asset-based location for the military as developed by Multispectral Solutions [93]. For chipsets we look at well-known UWB startups, such as Time Domain and Xtreme-Spectrum, which have developed functional chipsets.

9.1 MILITARY APPLICATIONS

As with many wireless communication technologies, the military has been the major driving force behind the development of UWB. In particular, radar applications have been developed by the military for many years. See, for example, [1] for a large number of UWB radar examples.

Ultra Wideband Signals and Systems in Communication Engineering   Second Edition
M. Ghavami, L. B. Michael and R. Kohno   © 2007 John Wiley & Sons, Ltd

9.1.1 Precision asset location system

Using the location and communications aspects of UWB, one extremely interesting military application is the *asset location system* developed by Multispectral Solutions and the US Navy [93]. This system, while military in design, has immediate commercial applications, since it is neither an offensive nor a defensive device, but rather a wireless system to improve logistics by knowing the location of containers and other large objects within a Navy ship at all times.

The major reason behind the development of the UWB precision asset location (PAL) system was the massive mobilization of military forces and goods in Desert Storm. It was reported that 40 000 containers were shipped; however, more than half, approximately 25 000 containers, had to be opened to check the contents due to inaccurate or lost paper invoices.

In Navy ships, narrowband radiofrequency identification (RFID) tags have not worked well due to excessive multipath (i.e. large delay spreads) and limited accuracy. To overcome these obstacles a UWB PAL system was developed.

The system consisted of UWB tags, which were placed on the devices whose location was to be measured and receivers, which were placed at fixed locations in the cargo hold where the object locations were to be measured. The UWB tags consisted of a short-pulse transmitter with a measured peak output power of approximately 250 mW. Each burst of information consisted of 40 bits and was repeated at 5 s intervals. The instantaneous bandwidth of the transmitted pulses was approximately 400 MHz.

In this system the short-pulse width property was used to make fine time-of-arrival measurements of the signals from each of the uniquely identified tags.

Tests were conducted on this system in a military ship's cargo hold, which measured approximately 25 m by 30 m and was 9 m high. Tests were performed both with no cargo and with cargo. The cargo consisted of 20 ft ISO containers stacked singly and doubly and HMMWV (high mobility multipurpose wheeled vehicle) trucks. Tags were located on the top of the containers and HMMWVs.

The accuracy of the system was between 1–2 m rms for single-stacked containers and approximately 4 m for double-stacked ones. Depending on the particular tag and test, an accuracy of less than 1 m was possible.

Further tests to determine the multipath environment showed that the typical delay spread was 3 μs, which is typically an order of magnitude greater than other indoor environments, such as offices. Frequency domain multipath nulls between 30 dB and 40 dB were also measured.

The PAL system shows the promise of UWB communications and ranging applications because short pulses can give extremely accurate results, even in extreme multipath environments. It was concluded for this particular application that UWB tags outperformed the current narrowband RFID tags.

Example 9.1

For the PAL system described in this chapter, calculate the average transmitter power of the system (in nW). Assume that transmitting one bit requires 2.5 ns. Describe the output power of the system in W/MHz and dBm/MHz. Is this system above or below FCC Part 15 limits?

Solution

The peak power of the transmitter is 0.25 W, and the 40 bits of information are retransmitted after a 5 s break. The total time T required for 40 bits of information is

$$T = 40 \text{ bits} \times 2.5 \text{ ns} = 100 \text{ ns} \tag{9.1}$$

The amount of time required for transmission T_t is only 100 ns/5 s (i.e. 0.000 002%). The average transmit power P_{av} is calculated as

$$P_{av} \text{ [W]} = P_{peak}T_t = 0.25 \times 20.0 \times 10^{-9} = 5.0 \times 10^{-9} = 5 \text{ [nW]} \tag{9.2}$$

Since the system has a bandwidth of approximately 400 MHz, the output power of the system can be described as

$$P_{av} \text{ [W/MHz]} = \frac{P_{av}}{400} = 12.5 \times 10^{-12} \text{ [W/MHz]} \tag{9.3}$$

When converted to decibels

$$P_{av} \text{ [dBm/MHz]} = 10 \times \log{(P_{av})} + 30$$
$$= -109.03 + 30$$
$$= -79.03 \text{ [dBm/MHz]} \tag{9.4}$$

Note here that the 30 dB added was to convert from dB to dBm. Since FCC regulations for Part 15 emissions are set at -41 dBm/MHz, the power output of the precision location system is approximately 38 dB below FCC limits.

9.2 COMMERCIAL APPLICATIONS

At the time of writing of this book the number of applications using UWB technology which have passed beyond the prototype stage to commercialization are limited. Consumer devices are even further in the future. Most UWB devices in actual use are confined to the government or military, and many are UWB radar devices, which in this book have taken a minor role. However, many consumer electronic companies are actively developing chipsets and wireless applications which will use a UWB physical layer.

In this section we list some of these companies and introduce their vision for commercial UWB applications.

9.2.1 Time Domain

Time Domain develops customized PulsON chipsets for applications and solutions for its partners and industries. They are working with strategic partners to integrate customized PulsON chipsets into a variety of products to bring substantial new benefits to consumers, businesses, and government users. Devices using PulsON technology will possess inherent advantages such as ultra-high-speed data transmission, ultra-low power consumption, exceptional immunity to interference from other radio signals, and no interference with existing users.

9.2.1.1 Time Domain PulsON 200   Time Domain [4] has developed a UWB chipset commercialized under the PulsON brand. The first-generation PulsON 100, the later PulsON 200, and future chipsets form their product range. Time Domain is the first company to pass the FCC certification procedure for a communications product. The product is known as the PulsON 200 Evaluation Kit. It is a general platform to help end developers make consumer products, including wireless communications, tracking, and radar. A photograph of the wireless radios developed for the PulsON 200 Evaluation Kit are shown in Figure 9.1.

Fig. 9.1   PulsON 200 Evaluation Kit UWB radios. Reproduced by permission of ©Time Domain.

The PulsON 200 Evaluation Kit has a maximum raw data rate of 9.6 Mbps over greater than 10 m in an environment free of obstacles and approximately 7 m in a residential or office environment. The detailed specifications of the PulsON 200 are shown in Table 9.1.

Future versions of PulsON are under development specifically for very high-bandwidth and low-power-consumption devices with the primary application of wireless multimedia.

9.2.1.2 Time Domain UWB signal generator   Time Domain has also developed a signal generator which functions as a UWB transmitter. It can be used in conjunction with a digital oscilloscope to characterize the UWB channel. It can also be used

Table 9.1Time Domain's PulsON 200 Evaluation Kit specifications.

Parameter	Value
Pulse repetition frequency	9.6 MHz
Data rates	9.6, 4.8, 2.4, 1.2, 0.6, 0.3, 0.15, 0.075 Mbps
Center frequency	4.7 GHz
Bandwidth	3.2 GHz (10 dB radiated)
EIRP	−11.5 dBm
Power consumption	12.2 W (transmit)
	11.9 W (receive)
FCC compliance	Parts 15.517, 15.209
Modulation	BPM, quadrature flip-time modulation

with other UWB devices to examine the effects of interference and coexistence. A photograph of the PulsON 200 signal generator is shown in Figure 9.2.

Fig. 9.2 PulsON 200 UWB signal generator. Reproduced by permission of ©Time Domain.

9.2.1.3 Time Domain PulsON 210 The current generation of PulsON products, PulsON 210, supports customers, beginning with UWB familiarization through every step to final UWB-based product development. Time Domain's PulseOn 210 Reference Design (P210 RD) is used for developing demonstrations or prototyping products. It is driven by P210 Integratable Module (IM), a highly flexible, easy to integrate UWB engine. The P210 RD adds an example customer interface module to the IM, encloses the electronics in a robust housing, and comes with a comprehensive software and documentation package. With the detailed documentation, the P210 RD serves not only as a prototyping tool, but includes a hands-on tutorial on the integration of the P210 IM.

9.2.2   XtremeSpectrum

XtremeSpectrum [5] has developed a four-chip lineup for UWB applications. The chipset is named XS100 TRINITY. The four chips are the RF front end (XSI102), RF transceiver (XSI112), digital baseband (XSI122), and MAC (XSI141). The chips were developed using low-cost 0.18 μm CMOS and SiGe (silicon-germanium) technology.

The RF front end consists mainly of a low-noise amplifier to boost the received UWB signal. High-gain (20 dB) and low-gain (0 dB) modes are provided. The noise figure is 5.6 dB. The RF transceiver consists of transmitter and receiver circuitry, and timing and bias/control circuitry. The digital baseband chip consists of an ADC, baseband circuitry, and the interface to the MAC chip. A flat, planar design omni-directional antenna 2.5 cm square has also been developed.

The detailed specification of the TRINITY chipset is shown in Table 9.2.

Table 9.2   XtremeSpectrum TRINITY chipset specifications.

Parameter	Value
Frequency band	3.1–10.6 [GHz]
MAC protocol	IEEE 802.15.3
Range	10 m
Modulation	BPM
Pulse type	Monocycle
Data rates	25, 50, 75, 100 Mbps
Power consumption	200 mW
Output power	<1 mW
BER	10^{-9}
Coding rates	1, 3/4, 1/2
Network	Peer-to-peer, ad hoc, piconet

In 2003 Motorola bought XtremeSpectrum, in a move that could let it jump to market with super-fast wireless broadband technology.

9.2.3   Intel Corporation

The Intel Corporation [97] has been involved in UWB both in the standards and research areas. Intel's contributions to technical advancement and general industry knowledge have established Intel research and development (R&D) as a leader in UWB technology. In the company it is believed that UWB is well suited for high-speed, short-range, wireless personal area connectivity for PC and mobile devices. Intel's current efforts focus on the three key hurdles: increased knowledge of UWB for the target usage models; creation of open standards for high-speed, short-range communications; and the need for worldwide regulatory approval. Intel's researchers are applying their expertise in CMOS radio design to understand the optimum requirements for low-power, low-cost UWB radios. Intel expected initial market deployment of standards-based UWB solutions to be sometime in the 2005–2006 time frame.

Intel's vision of UWB radio is of a 'common UWB radio platform' spanning many different applications and industries. UWB radio, along with the convergence layer, becomes the underlying transport mechanism for different applications. Some of the more notable applications that could potentially operate on top of the common UWB radio platform would be universal serial bus (USB), IEEE 1394/FireWire, and the next generation of Bluetooth.

In June 2003, Intel helped form the MultiBand OFDM Alliance (MBOA), with many of the most influential players in the consumer electronics, personal computing, home entertainment, semiconductor, and digital imaging market segments. In 2005, the MBOA merged with the WiMedia Alliance. Operating as the WiMedia Alliance, Inc., the combined organization continues to drive the standardization and adoption of UWB for high-speed wireless, multimedia-capable personal area connectivity.

The goal of this organization is to develop the best technical solution for the emerging UWB (IEEE 802.15.3a) Physical and MAC layer specification for a diverse set of applications.

9.2.4 Motorola

Motorola's Semiconductor Product Sector [98] has recently announced that it will use the company's UWB technology in its own products. It also backed XtremeSpectrum's UWB proposal at the IEEE 802.15 working group.

The two companies have signed a memorandum of understanding to work together to bring UWB technology to market and, eventually, deliver joint products to the marketplace.

Motorola is looking at various alternatives in the UWB market, but has decided to partner with XtremeSpectrum in part because they already have a working silicon solution.

9.2.5 Freescale

Freescale Semiconductor, Inc. is a global leader in the design and manufacture of embedded semiconductors for wireless, networking, automotive, consumer, and industrial markets. The company provides equipment manufacturers with chips to help them drive advanced cell phones, manage Internet traffic, and to help make vehicles safer and more energy efficient.

The Freescale XS110 UWB Solution, which is commercially available, has the capability of achieving over 110 Mbps data rates, while consuming minimal power. This helps to bring seamless, low-cost, highly reliable digital connectivity to consumers' homes and to the palms of their hands.

The XS110 UWB solution provides full wireless connectivity implementing direct sequence UWB (DS-UWB) and the IEEE 802.15.3 MAC protocol. The chipset supports applications such as streaming video, streaming audio, and high-rate data transfer at very low levels of power consumption. In addition to high data rates, the XS110 supports peer-to-peer as well as ad-hoc networking for truly mobile wireless connectivity. The XS110 is a wireless solution when high data rates, low power, and low cost are mandatory.

The Freescale DS-UWB implementation delivers wireless without compromise and is currently the only commercially available UWB chipset that has received FCC certification.

9.2.6 Communication Research Laboratory

In Japan the Communication Research Laboratory (CRL) has recently established a project group devoted to UWB in order to promote the R&D of UWB technologies. It has been investigating appropriate specifications for radio regulation on UWB systems [99]. The project focuses on the R&D of UWB systems in the microwave and millimeter wavebands as a result of industrial demands and for academic novelty in research.

The project group consists of a collaboration among industry, academia, and government. The CRL then established a UWB consortium together with industrial companies and universities with the support of the Yokosuka Research Park (YRP) in September 2002. The aims of this UWB Consortium are described as follows:

- R&D of all technologies for UWB wireless access systems;

- implementation and experimental investigation of a test bed using a microwave and submillimeter waveband system (i.e. 960 MHz, 3.1–10.6 GHz, and 22–29 GHz);

- R&D of UWB systems in unused high-frequency bands, such as the millimeter waveband over 60 GHz;

- establishment of transmission systems based on UWB, with low-cost and high-speed transmission over 100 Mbps;

- contribution to the standardization of UWB systems both in Japan and overseas.

9.2.7 General atomics

General Atomics [100] is pursing a communication product based on UWB multiband technologies. The data rate of the chipset is expected to be 120 Mbps. The target products are wireless transmission of multiple video streams and high-speed cable replacement.

9.2.8 Wisair

Wisair [101] is a startup company based in Israel which has developed a UWB chipset aimed at wireless indoor communications. Simultaneous multistreaming of audio and video, broadband multimedia, and quality of service are targeted, to support a wide range of entertainment and communication applications. The name of the chipset is UBLink.

The most interesting technical feature of the Wisair chipset is that it uses the *multi-band* approach, dividing the UWB wireless channel spectrum into several narrower bands. In this chipset, there are 30 possible sub-bands, of which one to 15 can be used. This approach provides flexibility of channels to use in a particular environment. To combat noise at a particular frequency, only a select subset of sub-bands can be used, rejecting low-quality sub-bands. For low-bit-rate applications, again a lower number of sub-bands can be used to save power or reliability can be increased by changing the coding rate. However, the disadvantage of this approach is complexity and cost. It will be interesting to see which UWB technologies provide the most benefit at the lowest cost.

Since the focus for this chipset is the consumer product market, low power consumption is a primary target. Thus, a power-saving standby mode has been developed as well. Details of the UBLink chipset are shown in Table 9.3.

Table 9.3 Parameters of the Wisair UBLink chipset.

Parameter	Value
Bit rates	20, 62.5, 83.3, 125 Mbps
Range	10–30 m
Power consumption	60–200 mW
Multiband	1–15 sub-bands

9.2.9 Artimi

Artimi is a semiconductor company developing a single-chip WiMedia-based system for low-power, high-bandwidth wireless connectivity based on UWB technologies. Artimi's products are system solutions which include radio, baseband, MAC, I/O, and software. According to [102], they are ideal for high-speed wireless USB applications.

Artimi achieved a major milestone in May 2005 with the release of the RTMI-100, the first complete single-chip UWB system solution including radio, physical layer and MAC. Artimi's next product, the RTMI-150, is a low-power, highly integrated, complete UWB system on a chip (SoC) providing a battery optimized wireless replacement for high-speed data transfer, synchronization, and streaming media in power sensitive consumer, communication, and PC peripheral devices such as digital still cameras, camcorders, MP3 players, and mobile phones.

9.2.10   Ubisense

Ubisense is a new company selling a robust, scalable platform that uses UWB technology to enable fine-grain tracking of people and objects in buildings and built-up areas [103]. Ubisense is headquartered in Cambridge, England.

Ubisense technology is composed of a real-time software platform, a network of UWB sensors, and a series of tags worn by staff or attached to objects within the workspace. Unlike systems based on conventional RF technology, which can have problems with accuracy and penetration of walls, the Ubisense system relies on short-duration UWB pulses that can locate, in real time, staff and equipment to within 15 cm.

9.2.11   Home networking and home electronics

One of the most promising commercial application areas for UWB technology is the wireless connectivity of different home electronic systems. It is thought that many electronics manufacturers are investigating UWB as the wireless means to connect devices, such as televisions, DVD players, camcorders, and audio systems, together to remove some of the wiring clutter in the living room. This is particularly important when we consider the bit rate needed for high-definition television that is in excess of 30 Mbps over a distance of at least a few meters. An example of a possible home-networking setup using high-speed wireless data transfer of UWB is shown in Figure 9.3.

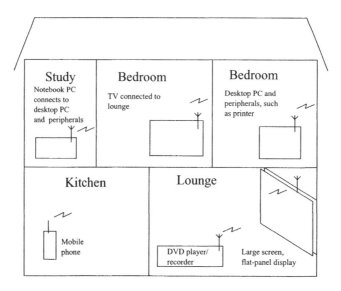

Fig. 9.3   An example of a possible home-networking setup using UWB.

Of course, UWB wireless connections to and from personal computers are also another possible consumer market area, with products expected in the next few years.

In [104] a proposal is made to use UWB as the wireless link in a ubiquitous 'homelink', which consists of an amalgamation of wired and wireless technologies. The wired technology proposed by the authors is based on the IEEE 1394 standard. This is an attempt to effectively integrate entertainment, consumer communications, and computing within the home environment. The reason for the choice of IEEE 1394 is that it provides an *isochronous mode*, in which data are guaranteed to be delivered within a certain time frame after transmission has started. Bandwidth is reserved in advance, which gives a constant transmission speed. This is important for real-time applications, such as video broadcasts, to ensure that there is no break in the movie or television program for the viewer. Some possible services and required data rates are shown in Table 9.4.

Table 9.4 Some possible contents for a home entertainment and computing network, the necessary data rates, and the possible need for real-time features.

Service	Data rate [Mbps]	Real-time feature
Digital video	32	Yes
DVD, TV	2–15	Yes
Audio	1.5	Yes
Internet access	>10	No
PC	32	No
Other	<1	No

Furthermore, IEEE 1394 has an *asynchronous* mode, in which data are guaranteed to be delivered, but bandwidth is not reserved and no guarantee is made about the time of arrival of the data.

IEEE 1394 provides scalable performance with 100, 200, and 400 Mbps, which is comparable with the target for UWB transmission speeds. IEEE 1394b is under consideration and will support 800 to 1600 Mbps, and may be extended to 3200 Mbps.

The considerations for home appliances can be described as economy, easy operation, flexibility, and high reliability. In the areas of economy (that is cheap devices) and reliability, in particular, UWB can be expected to perform well. However, there are various other wireless systems that are targeting this application. In particular, the established IEEE 802.11a standard [105], which has data rates of up to 54 Mbps, is a strong contender. IEEE 802.11a chipsets have been reported to have speeds of greater than 70 Mbps and are expected to increase past 100 Mbps in the near future. Alternatively, wireless transceivers based on the 802.11g standard may also provide an economical, if slightly lower, data rate.

Bluetooth [106], HomeRF [107], and 802.11b standards are not strong contenders, because their maximum data rates are approximately 720 kbps, 1.6 Mbps, and 11 Mbps, respectively. Another possible alternative is Wireless 1394, a wireless extension to the wired 1394 system.

9.2.12 PAL system

In the previous section we looked at the military use of the PAL system based on UWB devices, developed by Multispectral Solutions, Inc. In [108] the further commercialization and testing of the PAL system was reported. In particular, the PAL system was extended from military ships to use in hospitals and factories. The product was named PAL650 and was certified according to the FCC rules.

It was delivered as a commercial product at the time of writing to such clients as the Washington National Medical Center. The UWB precision asset location system enables hospital administrators to track the utilization of assets and patient flow accurately within the hospital. Tracking of patients and equipment is expected to increase hospital efficiency, particularly in the case of a terrorist attack or mass disasters, where hospitals see four to five times the normal load of patients.

Another client is the National Institute of Standards and Technology (NIST) in the US for precision tracking of robotic vehicles for search and rescue. The system will initially be used to evaluate robotic vehicle performance during international search and rescue tournaments. Future use will involve the tracking of rapidly deployable search and rescue vehicles as they enter buildings, following a disaster.

One technical item of interest for PAL systems is that they do not require a high data rate and, thus, may use significantly higher peak powers than is allowed for high-data-rate communication systems. The peak measured emission of the PAL650 system was 58.13 dBμV/m with an average emission of 39.01 dBμV/m. All measurements were performed with a 1 MHz resolution bandwidth (RBW) and referenced to a 3 m range. It is interesting to note that average emission measurements were limited by the testing equipment noise floor, and the true average EIRP for PAL650 can be calculated to be 16.1 dBμV/m. This value is approximately 37 dB below the FCC limit and over 22 dB below the noise floor.

The commercialization of PAL650 leads to the requirement of long battery life. Operation lifetimes in excess of 3.8 years are expected for these tags using a single Lithium cell CR2477 (3.0 V, 1 Ah). The tag operates at 3.0 V with a current consumption of approximately 30 μA.

As with the previously described PAL system, PAL650 consists of a set of active tags, UWB receivers, and a central processing hub that processes the received signals from the active tags.

A photograph of an active tag with the polyethylene radome cover is shown in Figure 9.4. The size of the tag is approximately 4.75 cm in diameter and 2.25 cm high when not covered in the housing. With the radome housing the size expands to 5.1 cm and 2.85 cm. The UWB active tag has a center frequency of 6.191 GHz and an instantaneous −10 dB bandwidth of 1.25 GHz.

A photograph of the receiver is shown in Figure 9.5. The size of the receiver boards are approximately 5.5 cm by 9.0 cm and 2.5 cm high. When the housing for the receiver unit is included, the total size becomes 5 cm by 8.25 cm by 15.25 cm. The low-noise receiver front end is housed in the gray box in the top left of Figure 9.5.

Fig. 9.4   PAL650 UWB active tag with radome. Reproduced by permission of ©2003 IEEE.

Receivers obtain power via the central processing unit, and data are transferred by a wired system. The wireless part is between the receivers and the active tags. The system described in [108] is a daisy-chained one; however, a hub-and-spoke system is under development to eliminate the potential for a single point of failure in the serial communications and power distribution to render the system unusable. The hub-and-spoke system will increase the reliability of the system as a whole.

A photograph of the central processing hub is shown in Figure 9.6. The central processing unit receives signals from the receivers and calculates the position of the active tags.

Calibration is performed at startup using a reference tag, whose location is known in advance. The current system receives updates from tags once per second in a burst of 72 pulse (bits) at a 1 Mbps burst rate. These data include synchronization preamble, tag identification, data field, forward error correction, and control bits. Update rates of up to 5200 per second can be accommodated without exceeding present FCC limits.

9.3   UWB POTENTIALS IN MEDICINE

UWB radar has been known to be a possible wireless tool to detect respiratory and cardiac function in patients. This application was first presented in the 1970s but was rejected for a number of reasons including safety, size, cost, and convenience.

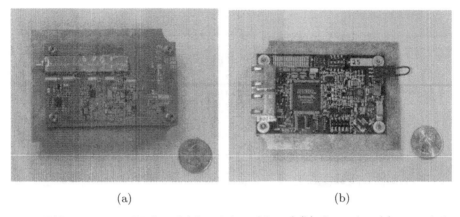

(a)   (b)

Fig. 9.5 PAL650 receiver RF board (a) and digital board (b). Reproduced by permission of ⓒ2003 IEEE.

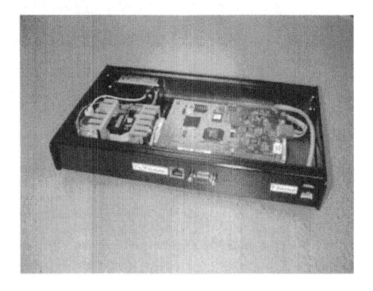

Fig. 9.6 PAL650 central processing hub. Reproduced by permission of ⓒ2003 IEEE.

However, the interest has recently been renewed with the advances of UWB. It is also suggested that UWB can be used as a form of medical imaging in much the same way as ground-penetrating radar provides images of objects in the ground. There have been a variety of suggestions for applications in the medical field [109]:

- *Cardiac biomechanics assessment*: This is the investigation of the cardiac function in both health and disease with special attention to the role of mechanics.

- *Chest movements assessment*: Observation of chest wall movement using UWB techniques is an excellent source of information about respiratory diseases.

- *Obstructive sleep apnea monitors*: This is the observation of the situation where the throat is sucked closed during sleep. This may happen in 4% of middle-aged people.

- *Soft-tissue biomechanics research*: Recently, research is growing on the biomechanics of soft tissues such as tendon, ligament, and cartilage. This interest is spurred by the need for more realistic medical simulations.

- *Heart imaging*: Heart scans are used to examine the flow of blood to the heart muscle and how well the heart pumps the blood.

- *Chest imaging*: Chest imaging remains a major component of diagnostic radiology. Plain chest radiography and chest computed tomography (CT) comprise about 20% of all procedures performed in the radiology departments.

- *Cardiac monitoring*: The cardiac monitor continuously displays the cardiac parameters such as cardiovascular pressures for patient diagnosis and treatment.

- *Respiratory monitoring*: This is the observation of changes in breathing patterns particularly during sleep.

- *Sudden infant death syndrome (SIDS) monitors*: SIDS occurs very rapidly, usually during sleep, and is the leading killer of infants between 1 week and 1 year with an approximate rate of two per 1000 live births.

- *Vocal tract studying*: In the field of speech research there has long been a desire to obtain a body of morphological information about the shape of the vocal tract as well as other airways, such as the nasal tract and trachea, to aid in further understanding of voice and speech production. Modern imaging techniques are beginning to be used to acquire three-dimensional shape information about the vocal tract and associated airways.

With the combination of an infrared laser diode which emits a short packet of electromagnetic waves as the transmitter and a UWB receiver equipped with a photodiode, possible applications include:

- noninvasive biochemical study of soft tissues;

- noninvasive study of metabolic processes;

- infrared spectral imaging.

9.3.1 Fundamentals of medical UWB radar

Short-duration electromagnetic pulses transmitted from a UWB radar are able to penetrate and possibly probe the human body. For instance, a UWB radar system may be able to detect, noninvasively, the movements of the heart wall. This is because of the difference in reflection magnitude between the heart muscle and the blood it pushes into the vascular tree.

In a patent [110] awarded to McEwan on medical UWB radar, a rough description of the physical principle of operation is given. As the impedance of the cardiac muscle is of the order of 60 Ω and the impedance of blood is about 50 Ω, a roughly 10% reflection magnitude of the RF energy at the heart muscle/blood boundary can be expected.

Let μ_0 and ϵ_0 be, the permeability and permittivity of vacuum, respectively; then the impedance of the heart muscle is

$$Z_{\text{heart}} = \sqrt{\frac{\mu_0}{\epsilon_r \epsilon_0}} = 60 \quad [\Omega] \tag{9.5}$$

where ϵ_r is the relative permittivity of the muscle, approximately equal to 40, while the impedance of blood, with the relative permittivity of 60, is 49 Ω. The reflection coefficient, defined as $(Y-1)/(Y+1)$, where

$$Y = \frac{Z_{\text{heart}}}{Z_{\text{blood}}} \tag{9.6}$$

gives a 9.9% return fraction of the radiated pulse. The same rule should be applied for reflection at the air/chest interface or the chest/lung interface, and even at vessels boundaries.

However, this model [110] does not account for the various living tissues through which the incoming pulse has to travel before reaching the useful echo surface on the heart wall.

9.3.2 UWB radar for remote monitoring of patient's vital activities

The possibility of utilization of UWB radar systems for remote measurement of a patient's heart activity and respiration is reported in [111]. UWB signals with a duration of 0.2 to 1 ns are used to transmit electromagnetic waves which are partially reflected from the interface of two media.

The method used is quite simple. The UWB transmitter generates the sequence of short pulses with a duration of about 250 ps, which excite an antenna and are radiated in space.

If d is the spatial distance between pulses and c is the speed of light, the pulse repetition cycle will be equal to

$$T = \frac{d}{c} \tag{9.7}$$

The pulse repetition frequency is defined as

$$f = \frac{1}{T} = \frac{c}{d} \tag{9.8}$$

This pulse sequence is reflected from the object. The pulse repetition cycle remains constant if the object is stationary.

When the object is moving with a harmonic rule, the velocity of its motion can be written as follows:

$$V = V_s \sin(\omega_s t) \tag{9.9}$$

Thus, spatial distance between pulses is written as

$$d_s = d - TV_s \sin(\omega_s t) \tag{9.10}$$

and the pulse repetition cycle is equal to

$$T_s = \frac{d - TV_s \sin(\omega_s t)}{c} \tag{9.11}$$

Consequently, the pulse repetition frequency will be

$$f_s = \frac{1}{T_s} = \frac{c}{d - TV_s \sin(\omega_s t)} = \frac{f}{1 - (V_s \sin(\omega_s t)/c)} \tag{9.12}$$

As can be observed in Equation (9.12), we obtain a frequency-modulated signal with nonlinear dependence between the frequency and the speed of motion.

The modulated sequence of pulses, carrying information about the parameters and the characteristics of organs under examination, is then digitized and is entered into a computer for further off-line signal processing and selection of the required information.

9.3.3   UWB respiratory monitoring system

One of the promising biomedical applications of UWB technology is respiratory monitoring. By transmitting pulses whose duration is of the order of a nanosecond, and monitoring relative time differences in signals reflected from the thorax during inhalation and exhalation, the displacement of a human chest can be monitored very accurately.

Currently, respiration patterns are measured by direct wire techniques, which may be inconvenient, particularly for infants, as they require multiple input and sensor lines attached to the patient. By using UWB signaling, not only could one avoid direct skin contact with the patient, but one could also achieve time resolution of the order of a fraction of a nanosecond, which is presently not possible using any other techniques. Such systems for respiratory monitoring are particularly important in preventive medicine, such as, for example, to alert for sudden infant death syndrome and in medical diagnosis. Many respiratory diseases (such as pneumonia, lung cancer, phthisic, etc.) induce the so-called pulmonary fibrosis, and, as a result, the respiratory frequency of a patient with the disease often increases. There are also other flu-like diseases that can greatly affect the pulmonary function.

Respiratory monitoring devices are now playing an important role in medical applications, and making them safe, cheap, and relatively simple (for home use) is of great importance. Furthermore, as the UWB signals have the capability of propagating through a wide range of materials (i.e. concrete, ice, ground), one can also envisage other applications. Rescue operations (e.g., detecting respiratory movements of humans under ruins) and home health care are some of the potential areas of application.

9.3.3.1 A simple finite-difference time-domain technique for detection of respiratory movements Consider a system where the transmitter periodically transmits a sequence of UWB impulses in a low duty-cycle mode. Such an assumption on the transmitted signal is made because only the time difference between the reflected signals for fully inflated and fully deflated lungs should be sufficient, and therefore it is not necessary to have a continuous transmission. Furthermore, this also provides low-power-consumption benefits. Another possible method of operation is to send bursts of pulses during each phase (inhalation or exhalation). Under such conditions, one can assume that the movement of a human chest within each transmission cycle (or several cycles) is negligible.

Assume that each pulse in a sequence is a Gaussian doublet $y_{g3}(t)$ defined in Equation (2.7). The transmitted signal can be expressed as

$$x(t) = \sum_k g(t)y_{g3}(t - kT_c) \tag{9.13}$$

where T_c denotes the period of the cycle transmission and $g(t)$ is a known PN used for coding the stream of transmitted impulses.

A simplified model of a horizontal slice of the human chest can be thought of as having a multilayer structure, where the interfaces between different layers are assumed to be flat. It is also assumed that all tissues are linear, homogeneous, and isotropic media with parameters that can be obtained from [112]. It should be noted that the real layer interfaces are not ideally flat. However, if the cell size is small and the one-dimensional case is considered, the flat interface model can be a reasonably appropriate model [113].

At the receiver, a subspace-based approach can be used in conjunction with sub-Nyquist sampling to estimate the time delay of the signal reflected from the thorax [114]. It is assumed that every transmitted pulse is emitted within the time interval T, which is large enough to capture all interfering signals prior to the transmission of the next impulse in the sequence, as illustrated in Figure 9.7.

The time interval T can be divided into three sub-intervals: the round-trip time (RTT), the receiving window (RW), and the guard interval (GI). RTT and GI must be chosen such that they can capture most of the interfering signals. For instance, multipath echoes due to surrounding objects should arrive within RTT, whereas GI

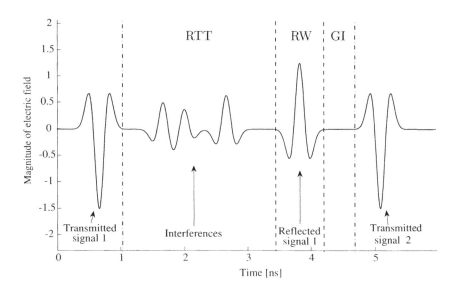

Fig. 9.7 Illustration of different time windows for transmitted, reflected, and interfering signals.

should capture a possible excitation that appears after the reflected signal, and which may affect the next transmitted impulse.

A mathematical technique called the finite-difference time-domain (FDTD) method represents an accurate approach for solving Maxwell's equations [115]. Recently, with the increased interest in applying the FDTD method in numerical simulations involving biological tissues, a frequency-dependent FDTD method has been proposed, which can be used for very accurate simulations of interactions between electromagnetic waves and a variety of dispersive materials [116]. There are different approaches for implementing this method. It is assumed that the UWB signal propagation resembles uniform plane wave propagation and the net free charge in the region is zero, regardless of the displacement current.

9.4 SUMMARY

In this chapter we investigated a brief selection of current UWB applications, focused on consumer communications. We presented material about current wireless UWB chipsets, by such companies as Time Domain, XtremeSpectrum, and Intel Corporation. As a test case we examined the PAL for the military as developed by Multispectral Solutions. We noted that consumer products are expected to appear within the next few years.

We also considered the potential of UWB in medicine. It was observed that UWB can be suggested for a variety of applications in the medical field such as remote monitoring of a patient's vital activities and a respiratory monitoring system.

Problems

Problem 1. Survey three current UWB products. Compare and contrast the specifications of each.

Problem 2. Design an original UWB product. What features does it possess to make it a success in the marketplace? Why must your product use UWB, as opposed to other wireless technologies?

Problem 3. Investigate how the ADC in the XtremeSpectrum chip achieves its very high sampling rate.

Problem 4. Investigate the specifications of PAL650. Is it a pulse-based UWB system?

10

UWB communication standards

A standard can be defined as a recommended practice in the manufacturing of a product or material or in the conduct of a business, art, or profession. Standards may be used as specifications from which to design a product and prove compliance. A standard plays a key role in creating industry efficiency, innovation, and interoperability between products created by different manufactures. Furthermore, standards have proven to be an important instrument for government deregulation in particular industries, as the existence of a standard provides a baseline for acceptance of a product. Government does not have to create new rules and laws for every single product that is produced, and standardized tests can be formulated. This leads to a profitable situation for both business and government as it allows for flexibility and innovation in individual product development, while adhering to certain common rules and guidelines.

10.1 UWB STANDARDIZATION IN WIRELESS PERSONAL AREA NETWORKS

An expected area for consumer communication devices using UWB is the area of wireless personal area networks (WPANs). Before we describe in detail the workings of the subgroups focused on UWB, we will provide a quick overview of all WPAN working groups to provide some context for the following discussion.

Ultra Wideband Signals and Systems in Communication Engineering   Second Edition
M. Ghavami, L. B. Michael and R. Kohno   © 2007 John Wiley & Sons, Ltd

10.1.1   WPAN standardization overview

The standards activity of WPANs takes place in IEEE 802.15, which is an international standards working group under the auspices of the Institute of Electrical and Electronics Engineers. The group involves dozens of major companies involved in UWB R&D and product and chipset design. The IEEE 802.15 group is responsible for creating a variety of WPAN standards, and is divided into four major task groups which are shown in Figure 10.1.

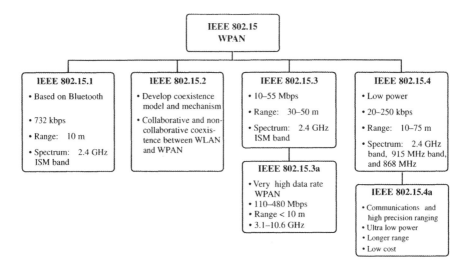

Fig. 10.1 Study groups of IEEE 802.15 and their respective focus for developing WPANs.

The IEEE 802.15.1 task group is working on Bluetooth v1.1. Bluetooth is a wireless protocol which uses a short-range radio link (up to 10 m) to transmit data between personal devices, forming an ad-hoc network in the unlicensed 2.4 GHz ISM band. The 802.15.1 standard will eventually allow data transfers between a WPAN device and a WLAN device. WLANs are defined by the IEEE 802.11 group.

The task group of IEEE 802.15.2 is concerned with coexistence issues that arise when two wireless systems share an environment of operation. The IEEE 802.15.2 task group had two goals: to quantify the effects of mutual interference between WPAN and WLAN devices, and to establish mechanisms for coexistence of WPAN and WLAN (e.g., IEEE 802.15.1 and IEEE 802.11b).

The IEEE 802.15.3 task group is developing WPANs with data rates up to 55 Mbps. The standard should operate on five 15 MHz channels in the unlicensed 2.4 GHz ISM band.

The IEEE 802.15.4 task group is focused on low-data-rate, low-power WPANs (LP-WPANs). IEEE 802.15.4 members are investigating low-data-rate WPAN solutions with a battery life ranging from months to several years, with very low device complexity.

Now we will proceed to examine the two study groups within 802.15 that are involved specifically with UWB standardization.

10.1.2 IEEE 802.15.3a

In 2001 an IEEE study group, called 802.15.3 SGa, was set up to study and propose an alternative to the official IEEE 802.15.3 standard. This was to serve the requirements of companies wishing to deploy very-high-data-rate applications, such as video transmission, with data rates greater than 110 Mbps at a distance of 10 m.

Probably the best candidate for definition of a new alternative was the UWB technology. The purpose of this study group was to provide a higher speed physical layer enhancement for applications involving imaging and multimedia. The main expected characteristics of this alternative can be summarized as follows [117]:

1. coexistence with all IEEE 802 wireless physical layers;

2. target data rate in excess of 100 Mbps for consumer applications;

3. robust multipath performance;

4. location awareness capability to enable applications such as range-dependent authentication and routing;

5. possible use of additional unlicensed spectrum for high-data-rate WPANs, relieving possible spectral congestion.

Moreover, it was anticipated that future applications would go beyond the 802.15.3 physical layer capabilities, such as, for example, with respect to higher data rates such as would be needed by multiple high-definition television (HDTV) channels and location awareness. Anticipated applications for 802.15.3a included:

1. multimedia (e.g., video, voice over internet protocol (IP), HDTV, home theater, surround sound audio, gaming);

2. location aware applications (e.g., location-dependent authorization);

3. digital imaging (faster and better resolution).

The technical requirements for the IEEE 802.15.3a proposal are summarized in Table 10.1.

The IEEE 802.15.3a study group was considered the main standardization organization for UWB. Since it began hearing proposals in March 2003, many companies have continuously improved and merged their ideas, and collaborated to form coalitions.

In June 2002, a UWB startup company, XtremeSpectrum, introduced the first UWB device under the new rules for wireless connectivity applications using its DS-UWB technology. XtremeSpectrum was acquired by Motorola in 2003 and has since become the UWB operation of Freescale Semiconductor, Inc. In 2004, Freescale

Table 10.1   Technical requirements for the 802.15.3a proposal.

Parameter	Value
Data rates	110, 220, 480 Mbps
Range	10 m, 4 m and below
Power consumption	100 mW and 250 mW
Power management modes	Capabilities such as power save, wake up, etc.
Co-located piconets	4
Interference susceptibility	Robust to IEEE systems
Coexistence capability	Reduced interference to IEEE systems
Cost	Similar to Bluetooth devices
Location awareness	Location information to be propagated to a suitable management entity
Scalability	Backwards compatibility with 802.15, adaptable to various regulatory regions (such as the USA, European countries or Japan)
Signal acquisition	Less than 20 µs
Antenna practicality	Size and form factor consistent with the original device

received FCC certification for the first UWB device for wireless communication applications under the UWB rules.

The UWB Multiband Coalition [118] was led by Intel Corporation and includes several other major companies that support a multiband approach which employs pulsed modulation. On July 2003, Intel and Texas Instruments (TI) merged their proposals to form a united approach that employs multiple bands and uses OFDM modulation. The newly formed Multiband OFDM Coalition [119], whose membership includes TI and the UWB Multiband Coalition, endorsed a proposal which is essentially the same as the original TI proposal, with an optional operating mode using seven bands.

After its July 2003 meeting, the IEEE 802.15.3a task group was left with two primary contenders:

1. the TI OFDM-based multiband approach which uses 528 MHz channels and was supported by the Multiband-OFDM Coalition; and

2. the XtremeSpectrum–Motorola dual-band impulse radio spread spectrum approach, which exploits all of the UWB spectrum allocation.

For more than three years, the IEEE study group 802.15.3a worked on the IEEE 802.15 UWB standard, striving to reach consensus between these two leading industry groups. Finally, it was agreed that such a consensus would not be reached, and the task group voted to recommend disbanding on January 2006 at the IEEE 802 meeting held in Hawaii. Nearly 95% of the members voted in favor of disbanding.

The recommendation was then forwarded to the 802.15 working group where it was approved. The dispute over which standard will become dominant has now passed to the marketplace.

10.1.3 IEEE 802.15.4a

The IEEE 802.15.4a task group is charged with developing an amendment to the current IEEE 802.15.4-2003 for an alternate physical layer. The primary interest in developing this alternate physical layer is to provide communications and high-precision ranging and location capability, high aggregate throughput, and ultra low power. The group is also looking at adding scalability to data rates, longer range, and lower power consumption and cost. These additional capabilities over the existing IEEE 802.15.4 standard are expected to enable significant new applications and market opportunities.

IEEE 802.15.4a became an official task group in March 2004. The committee is actively drafting an alternate physical layer specification for the applications identified in accordance with the project timetable.

In March 2005, the baseline specifications were selected without enacting any down-selection procedures, and, as a result, the baseline proposal was confirmed with 100% approval. The baseline proposal is two optional physical layers consisting of a UWB impulse radio (operating in the unlicensed UWB spectrum) and a chirp spread spectrum (operating in the unlicensed 2.4 GHz spectrum). The 802.15.4a compatible UWB impulse radio will be able to deliver communications and high-precision ranging [120].

10.2 DS-UWB PROPOSAL

Now we will move from a general overview of the history and work of UWB standardization to two particular examples which were proposed for the standardization of UWB in the WPAN group.

One of the proposals is the direct sequence UWB proposal, based on the original proposal by XtremeSpectrum (now Freescale Semiconductor, Inc.). The major proponent is the UWB Forum [121], a consortium of currently over 50 member companies. The proposed system uses DS CDMA with variable code lengths to provide data rates of 28 to 1320 Mbps over a spectrum of 1.8 GHz in the range from 3.1 to 4.9 GHz.

Figure 10.2 demonstrates a simplified block diagram of the baseband model of the DS-UWB transmitter. Information bits are first scrambled. Then, forward error correction (FEC) code is provided by convolutional codes with a coding rate of one-half. An additional coding rate is achieved by puncturing the encoder output. Encoded bits are then interleaved using a convolutional interleaver. Two types of modulation may be used, BPSK and 4-BOK (binary orthogonal keying). In the BPSK mode, 1 bit is used to determine the polarity of the spreading code which will be transmitted, while in the 4-BOK mode 2 bits are used to choose one of the two spreading codes as well as its polarity.

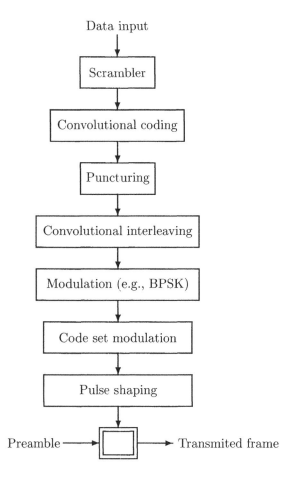

Fig. 10.2 A simplified model of a DS-UWB transmitter system.

Variable length codes combined with different chip rates as well as center frequencies are used to enable several devices to operate in the same band. Pulse shaping is then performed using a root-raised cosine low pass filter with 30% excess bandwidth.

10.2.1   DS-UWB operating bands

An update to the DS proposal describes the operating bands for the DS-UWB system. Each UWB communication device or piconet operates in one of two bands:

- 3.1–4.9 GHz (required);

- 6.2–9.7 GHz (for future use).

A piconet may represent a device such as a wireless printer used by a group of users in an office or simply two laptops exchanging files. Due to concerns over interference from 802.11a WLAN, the frequency range between 5 GHz and 6 GHz is avoided (Figure 10.3).

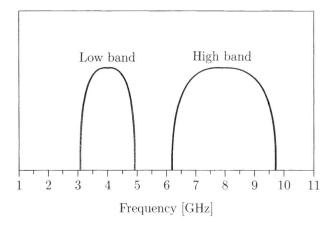

Fig. 10.3   DS-UWB bands.

One benefit of this approach is that it achieves a more even distribution of energy across the occupied band, compared to the MB-OFDM approach. Due to peaks and troughs of power across the band, the MB-OFDM approach creates higher peak-to-average power ratios.

The DS-UWB system will support a total of six simultaneous piconets within the 3.1 to 4.9 GHz band, each having 110 Mbps capacity. The understanding from industry discussion is that the technology may make use of the upper band in future, as silicon technology advances. Table 10.2 shows an overview of the DS-UWB proposal.

Table 10.2   Overview of the DS-UWB technique (subject to change).

Parameter	Value
Number of bands	2
Bandwidth	Approximately 50% fractional bandwidth
Frequency range	3.1–4.9 GHz and 6.2–9.7 GHz
Modulation schemes	BPSK, QPSK, DSSS
Coexistence method	Null band for WLAN ($\sim$5 GHz)
Multiple access method	CDMA
Number of simultaneous piconets	8
Error correction codes	Convolutional code, Reed–Solomon code
Code rates	1/2, 3/4, and 1
Chip time	A fraction of 1 ns
Multiple mitigation method	Decision feedback equalizer and rake

10.2.2 Advantages of DS-UWB

According to the supporters of the DS-UWB proposal, the following items are the key features of this technique:

- based on true UWB principles it provides large fractional bandwidth signals in two different bands;

- it benefits from low fading due to the wide bandwidth of greater than 1.5 GHz;

- it provides the best relative performance at high data rates;

- it can provide an excellent combination of high performance and low complexity for WPAN applications;

- it supports scalability to ultra-low-power operation for short-range, very high rates using low-complexity implementations;

- its performance exceeds the selection criteria in all aspects;

- it has better performance and lower power than any other proposal considered by IEEE 802.15.3a task group;

- it has an excellent basis for operation under 'gated UWB' rules, where UWB signals can be on and off during a certain period of time.

10.3 MB-OFDM UWB PROPOSAL

The MB-OFDM UWB proposal, mainly based on an original submission by TI, is supported by the MBOA, a consortium of over 150 member companies [122]. The proposed system uses OFDM along with time-frequency codes (TFC) to provide data rates of 55 to 480 Mbps over three frequency bands of 3.168 to 4.752 GHz.

Figure 10.4 demonstrates the block diagram of the baseband model of the MB-OFDM transmitter. Data bits are first scrambled. The scrambled bits are then encoded using a convolutional encoder. Encoded bits are then interleaved in a block interleaver and mapped to modulated symbols, using QPSK modulation. Modulated symbols are then mapped to 100 OFDM sub-carriers, along with 12 pilot and 10 guard tones, using a 128 point IFFT. Pilot tones are used for channel estimation and carrier-phase tracking at the receiver, while the energy transmitted on the guard tones may be adjusted to relax the transmit and receive filter specifications.

TFCs are used to interleave OFDM symbols over three frequency bands, each 528 MHz wide. This is done to enable the system to utilize more than one band for the given OFDM waveform and to enable simultaneous operation of multiple users over the three frequency bands by assigning different TFC codes to different users.

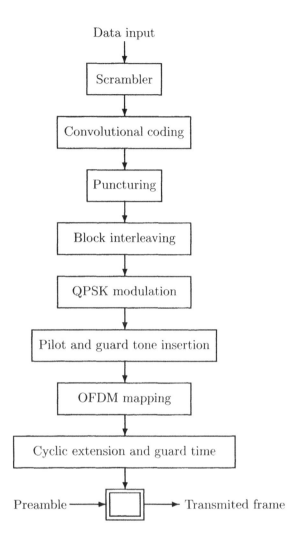

Fig. 10.4   A simplified model of an MBOA-UWB transmitter system.

10.3.1   Frequency band allocation

The relationship between the center frequency and band numbers is given by the following equation:

$$\text{Band center frequency} = 2904 + 528 \times n_b \qquad (10.1)$$

where $n_b = 1, \ldots, 14$ MHz. This definition provides a unique numbering system for all channels that have a spacing of 528 MHz and lie within the band 3.1–10.6 GHz. Based

on this, five band groups are defined, consisting of four groups of three bands each
and one group of two bands. Band group 1 is used for mode 1 devices (mandatory
mode). The remaining band groups are reserved for future use. The band allocation
is summarized in Table 10.3 and Figure 10.5.

Table 10.3 Frequency band allocation for the MB-OFDM UWB proposal.

Band group	Band number	Lower frequency [MHz]	Center frequency [MHz]	Upper frequency [MHz]
1	1	3168	3432	3696
	2	3696	3960	4224
	3	4224	4488	4752
2	4	4752	5016	5280
	5	5280	5544	5808
	6	5808	6072	6336
3	7	6336	6600	6864
	8	6864	7128	7392
	9	7392	7656	7920
4	10	7920	8184	8448
	11	8448	8712	8976
	12	8976	9240	9504
5	13	9504	9768	10 032
	14	10 032	10 296	10 560

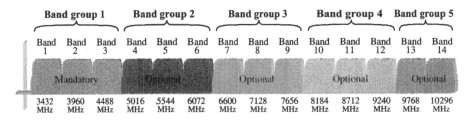

Fig. 10.5 Frequency band allocation for the MB-OFDM UWB proposal with the center
frequency in each band shown.

10.3.2 Channelization

By using a maximum of four different TFCs for each band group, a unique logical
channel can be dedicated to each piconet. The TFC patterns are defined in Table 10.4.

For example, for band group 2, the three band number values 4, 5, and 6 shall replace the values 1, 2, and 3, respectively, to generate the TFCs. For band groups 1 to 4, only TFCs numbered 1 to 4 with appropriate band number values shall be defined. For band group 5, only TFC numbers 5 and 6 shall be defined.

Table 10.4 TFC patterns for band group 1.

TFC number	Length 6 TFC (Band number values for band group 1)					
1	1	2	3	1	2	3
2	1	3	2	1	3	2
3	1	1	2	2	3	3
4	1	1	3	3	2	2
5	1	2	1	2	1	2
6	1	1	1	2	2	2

10.3.3 Advantages of MB-OFDM UWB

According to the supporters of the MB-OFDM UWB the popularity of the MBOA specifications and their adoption as the standard by these organizations is due to several advantages it provides over other approaches:

- *Spectral flexibility:* The MBOA will be able to ship a single solution worldwide because MB-OFDM can dynamically shape the spectrum via software controls.

- *Ability to be built in CMOS:* MB-OFDM technology was designed specifically to be built in low-cost CMOS semiconductor processes. An all-CMOS solution means that MB-OFDM will be much easier to integrate into a single-chip solution, as well as providing quick designs times and lower costs.

- *NBI management:* MB-OFDM can handle NBI completely in the digital domain with very simple and low-power techniques.

- *Control of out-of-band emissions:* Since the MB-OFDM solution is a digital solution, the out-of-band emissions are much more easily controlled.

10.4 A SHORT COMMENT ON THE TERM 'IMPULSE RADIO'

UWB traditionally meant impulse radio (IR), otherwise known as carrierless communication. Most wireless systems impress a signal on a carrier wave of a specified frequency. IR uses extremely short, highly synchronized electrical pulses, with no carrier wave involved. IR is what made UWB so novel, and gave it many of its beneficial properties.

The FCC's order did not require IR. It defined UWB as essentially any wireless system that spreads signals across at least 500 MHz of bandwidth. UWB is sometimes considered as the available spectrum instead of a specific technology.

IR can also be defined as a form of UWB signaling which is well designed for baseband asynchronous multiple access, short-distance high-data-rate multimedia services, and tactical wireless communications. IR, also referred to as single-band UWB, has also been considered as one of the two prominent approaches, so far investigated, to designing a UWB system. Most of the previous chapters of this book deal with the fundamental meaning of IR UWB systems.

10.5 SUMMARY

This chapter presented the standardization activities of IEEE 802.15 working groups on UWB systems. Among the technical requirements for the 802.15.3a proposal have been a data rate of over 110 Mbps and a maximum communication distance of 10 m. The IEEE 802.15.4a task group has been charged with developing an alternate physical layer to provide communications and high-precision ranging and location capability, high aggregate throughput, and ultra low power.

In this chapter we also presented an outline of the two physical layer proposals for 802.15.3a UWB systems, namely, DS-UWB and MB-OFDM. DS-UWB systems benefit from wider bandwidths and lower complexity, whereas the MB-OFDM technique has more spectral flexibility and a better ability to be built on CMOS.

Problems

Problem 1. What are the two main IEEE UWB standard bodies?

Problem 2. Compare IEEE 802.15.3a and IEEE 802.15.4a with regard to the range of communication and data rate.

Problem 3. How is dual band allocation in DS-UWB helpful for coexistence of UWB systems with other communication systems?

Problem 4. Is the MB-OFDM method meeting the basic definition of UWB systems regarding bandwidth?

11

Advanced topics in UWB communication systems

This chapter outlines some UWB communication systems that have been considered during the last few years. First, we will start with the application of UWB techniques to ad-hoc and sensor networks, including routing protocols. Then, space-time coding for UWB systems will be explained. Following that, self-interference in high-data-rate UWB communications will be discussed. Coexistence of DS-UWB with Wi-Max will be considered in more detail. Finally, vehicular radars in the 22–29 GHz band will be briefly explained.

11.1 UWB AD-HOC NETWORKS

11.1.1 Introduction

An ad-hoc network is a LAN or other small network in which some of the network devices are part of the network only for a short time, usually the duration of a communication session. Since the network forms on an ad-hoc basis, these networks are referred to as 'ad-hoc' networks. A particular feature of ad-hoc networks is that they are able to communicate effectively without base stations controlling the network, and communication occurs directly between devices.

Desirable features for ad-hoc networks are high precision location estimation abilities, extremely low power consumption, low production costs, autonomous operation, and adaptability to the changing wireless environment. UWB radio is considered an attractive transmission medium for ad-hoc networks. The combination of UWB technology and ad-hoc networks has recently been investigated [123–125].

Ultra Wideband Signals and Systems in Communication Engineering   Second Edition
M. Ghavami, L. B. Michael and R. Kohno   © 2007 John Wiley & Sons, Ltd

11.1.2   Applications of an UWB ad-hoc network

In an ad-hoc network each node has to have a routing function and it is essential to use multihop transmission to reach nodes further away. Since each node has a network control function, even if one of the nodes is not working properly, its influence on the whole network is quite limited. Therefore, ad-hoc networks are excellent with respect to cost and robustness. Hence, they have attracted much attention from researchers in both the military and the commercial fields. It is expected that UWB ad-hoc networks will be used for digital household electric appliances and peripheral equipment of PCs, for example, such as a wireless link between a PC and DVD player or a physical layer for a 'wireless USB' replacing traditional USB cables between devices.

As explained in Chapter 1, UWB systems can deliver more than 100 times the transmission rate of wireless protocols such as Bluetooth, while the power consumption of UWB systems is roughly equivalent. The maximum transmission data rate of IEEE 802.11a/g systems is about 56 Mbps, although the actual realizable rate of data transmission is usually less than 20 Mbps. In addition, construction of low-power devices (less than 1 W) poses considerable obstacles. For example, the certified transmission speed in wireless USB, which utilizes UWB in the physical layer, is specified to be more than 480 Mbps at a distance of 3 m. Wireless video transmission requires the transmission speed without compression to be more than 30 Mbps in order to transmit a signal with the same quality as HDTV. UWB can satisfy such high data rates. Combined with the technology of ad-hoc networking, UWB ad-hoc networks can be used for the following applications (Figure 11.1):

1. home theater systems (HDTV, DVD, and other high-definition media);

2. connection between peripherals of a PC (such as printers, digital or video cameras, music apparatus, PDAs);

3. connection from a mobile display to digital media servers (such as a DVD server to a PC display);

4. active RFID systems.

11.1.3   Technologies involved in UWB ad-hoc networks

Since the topology of an ad-hoc network is always changing, conventional static routing protocols cannot be employed. *Dynamic routing protocols* have already been widely investigated and deployed. The dynamic renewal of routing information enables the routing protocol to select the route for delivery of information adaptively. However, if a loop in the delivery route occurs, the load of the CPU and the traffic of the network are increased rapidly, and the network could then be brought down.

More recently, as the research into ad-hoc network advances, the use of positioning information has been considered in addition to the dynamic routing protocol. Thus, the use of UWB has attracted much attention. In a UWB ad-hoc network, the major research topic is the MAC. The MAC layer controls the delivery route, power control,

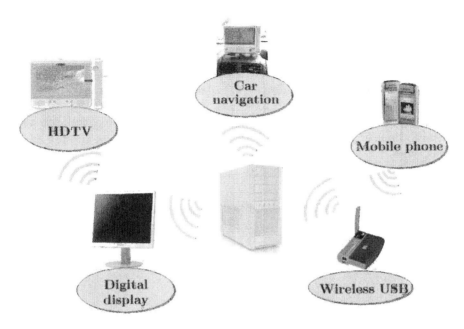

Fig. 11.1 Examples of applications for UWB ad-hoc networks.

and scheduling policies. In the next sub-section, routing protocols and automatic retransmission on request (ARQ) systems are introduced.

11.1.3.1 Routing protocols for UWB ad-hoc networks Routing protocols used in UWB can be classified into three groups: reactive, proactive, and hybrid systems [126]. These routing schemes do not employ location information about the nodes and rely on a flooding scheme. A flooding scheme is a simple mechanism and is highly reliable for the delivery of data. This scheme is not a good scheme for mobile nodes because the efficiency of bandwidth is not high. Routing schemes based on the SNR ratio have also been proposed.

Location-based routing systems can relay the packet to the desired region by the shortest route and therefore save network resources. Location-based routing systems with GPS have been considered. However, the resolution of GPS is not adequate for small mobile devices. Furthermore, it is difficult to use GPS indoors due to signal attenuation. UWB systems have been considered instead of GPS because the time resolution of UWB signals can be much higher then those of GPS. Therefore, investigation of the routing protocols for UWB is mainly concerned with location-based routing [127]. Furthermore, by using the positioning capabilities of the UWB MAC layer, its power consumption can be reduced, resulting in a highly efficient system.

11.1.3.2 ARQ protocols for UWB ad-hoc networks MAC protocols include the method for retransmission, power control, and scheduling between different terminals.

Retransmission is very important in order to compensate for the packet loss due to channel conditions and noise, and to improve the throughput. The ARQ protocol is usually used for retransmission. In the ARQ protocol, an acknowledge (ACK) packet is used for the confirmation of successful transmission and retransmission. When a receiver receives a packet successfully, an ACK packet is returned to the sender. When a packet is lost, which can be known because the packets are ordered, the receiver returns a negative acknowledge (NACK) packet to the sender and then the sender retransmits the lost packets (Figure 11.2). ARQ protocols can be divided into the following types:

1. stop-and-wait (SAW) ARQ;

2. go back N, (GBN) ARQ, where N is the number to go back;

3. selective-repeat (SR) ARQ;

4. hybrid ARQ.

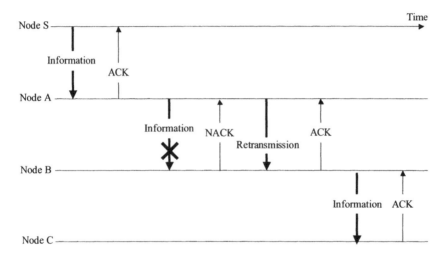

Fig. 11.2 An ARQ system in an UWB ad-hoc network.

All ARQ protocols use an ACK packet for confirmation. As the number of hops increases, more ACK/NACK packets are required in the system, increasing the total interference. It is therefore expected that a decrease in the number of ACK/NACK packets will improve the total network throughput. In this view, an echo detection scheme has been proposed. Echo detection is based on TDMA. This scheme checks in the time slot used by a relay node for the confirmation of successful transmission.

The code sense system has also been proposed as the optimal MAC protocol for DS-UWB systems [128]. This scheme uses the specific combinations of the spreading codes and does not need to use an ACK packet. These spreading codes are assigned to each node based on a node's ID and the other nodes can sense whether the target node is receiving a packet or not (Figure 11.3).

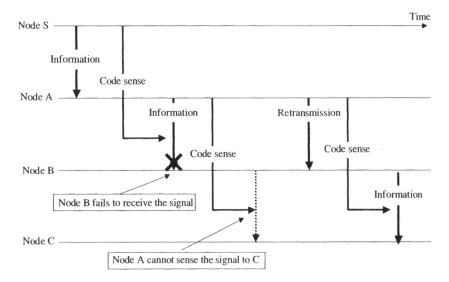

Fig. 11.3 A code sense system in an UWB ad-hoc network.

11.1.3.3 Summary Since UWB systems are quite different from conventional narrowband radio systems, MAC protocols optimized for UWB ad-hoc networks will also be different. In this section some of these protocols were introduced. A standard protocol has not yet been decided. Therefore, many universities and companies are continuing their investigation into MAC protocols of UWB ad-hoc networks.

11.2 UWB SENSOR NETWORKS

Wireless sensor networks are special networks comprising a large number of inexpensive and low-power sensor nodes, which are densely deployed to accomplish common tasks such as environmental monitoring, positioning, location tracking, and observation of the surrounding ambient conditions such as temperature, humidity, movement, and so on. Sensor nodes are battery powered and, due to their limited coverage range, they are deployed in an ad-hoc fashion and organize themselves to form a multihop wireless communication network (Figure 11.4).

Sensor networks have a wide variety of application areas such as the military, environment, health, home, and industrial areas. For example, networks of sensors at fixed positions could be installed in a conference room for person-to-person communication and interactive gaming, in a car for in-vehicle information exchange, or in a building for environmental control. A blanket of sensors could also be deployed in a

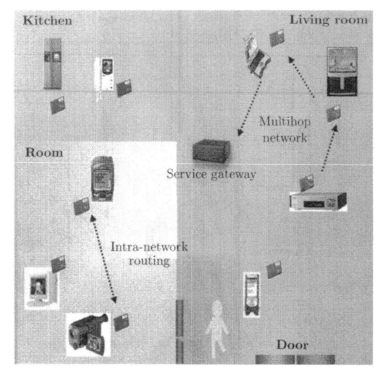

Fig. 11.4 An example of a UWB sensor network at home.

nonsecure military zone for battlefield surveillance, or could be dropped over a field for precision agriculture or environmental monitoring. Sensors could also be attached to the body to form a body area network (BAN) for either tracking and monitoring doctors and patients inside a hospital or for monitoring the movements and internal processes of insects and other small animals. Therefore, due to the large diversity of services offered by sensor networks in different applications it can be seen that sensor networks are expected to operate in many different environments ranging from indoor residential and office situations to larger industrial enclosures to outdoor suburban and agricultural environments to BANs.

In a wireless sensor network, transmissions of sensor nodes may be asynchronous and so there is an increased number of collisions resulting from multiple sensor nodes transmitting data simultaneously over the same medium. This is a severe problem facing communication in such networks, and can be said to be mainly due to the large number of nodes deployed. Thus, a reliable and efficient multiple access scheme is required to coordinate the transmissions. This is because many applications of sensor networks and, in particular, military applications require robust and rapid wireless networking in complex and hostile environments. Consequently, an ideal multiple access scheme for sensor networks should essentially allow increased immunity against fading and jamming (particularly in covert military networks), and a decreased level of interference, and should have increased energy saving capability to allow low-power communication, which is critical for sensor networks as the nodes are often required to last for several years.

UWB radio technology has a number of advantages that makes it a viable candidate for sensor networks. In particular, UWB systems have low power, low cost, and low complexity. They have a noise-like signal and are therefore resistant to severe multipath and jamming. They also provide greater bandwidth and are more cost-effective for providing increased throughput without increasing power.

The three approaches to multiple access in UWB are:

1. time hopping (TH-UWB);

2. direct sequence spread spectrum (DS-UWB);

3. orthogonal frequency division multiplexing (OFDM-UWB).

TH-UWB was introduced to eradicate collisions in UWB multi-user communication systems by time-hopping UWB pulses, representing a symbol by a pseudo-random code dedicated to a specific user. DS-UWB has the ability to mitigate interference, increase system capacity, and enhance the quality of service (QoS). It allows users to transmit in the same bandwidth simultaneously by pre-assigning a pseudo-random noise code that is multiplied to the transmit signal. OFDM-UWB allows multi-user communication by frequency hopping a UWB pulse train according to a pre-assigned frequency code sequence.

Extensive studies have been carried out by researchers on TH-UWB and DS-UWB systems, including multi-user performance analysis of TH-PPM, TH-BPSK, and DS-UWB systems, and comparisons of their performance. The performance analyses provided have been based on either a Gaussian approximation or a semi-analytical approach. Even though these contributions are useful they are not specific to wireless sensor networks. Since wireless sensor networks have certain stringent constraints that need to be satisfied, such as energy efficiency, scalability, throughput improvement, and effective collision avoidance, a further comparative analysis of these schemes is required [129].

The literature specifically aimed at the multiple access issues of UWB-based sensor networks is at present quite narrow. One study has been reported where the authors used simulations to compare the performance of TH-PPM and DS-UWB systems in a secure wireless sensor network in terms of link error probability. In addition, adaptation of multiple access parameters in a TH-UWB cluster-based wireless sensor network using a Gaussian approximation method has also been studied. In another study, a performance analysis of an UWB sensor network with the Aloha multiple access scheme has been presented, in which the performance was evaluated in terms of average link outage probability and the overall network throughput for different coverage radii of the sensor node. However, in all these works the effects of realistic application-specific propagation environments and an analysis of the primary performance factors of multiple access for wireless sensor networks are not assessed.

11.3 MULTIPLE INPUTS MULTIPLE OUTPUTS AND SPACE-TIME CODING FOR UWB SYSTEMS

UWB IR is defined as a form of ultra-wide bandwidth signaling which is well designed for baseband asynchronous multiple access, short-distance high-data-rate multimedia services, and tactical wireless communications. IR, also referred as single-band UWB, is one of the two prominent approaches so far investigated for the design of a UWB system. The other solution is called MB-OFDM.

The single-band UWB approach can be implemented by using simple pulse modulation such as PPM, PAM, PSM, or a combination of them. It is also involved with random TH or DS techniques to enable secure multiple user transmissions. The pulses are generated by one or several pulse shapers.

It is known that UWB signalling is particularly well designed to resolve dense multipath. In reality, the validity of this statement depends on the capacity of a rake receiver to capture and process a great deal of these multipaths, in the midst of strong inter-pulse interferences (IPI) resulting from the propagation characteristics. With this respect, the space-time coding (STC) technique can provide an interesting alternative against the sheer rake complexity. Moreover, multiple inputs multiple outputs (MIMO) systems are known to provide higher capacity and therefore better performance compared with a single link in wireless communication systems by employing multiple transmit and, optionally, multiple receive antennas.

So far, several schemes have been proposed to exploit the potential of MIMO architectures. Most popular are space-time block codes (STBCs) [130], space-time trellis codes (STTCs) [131], or layered space-time architectures (such as BLAST) [132]. All of these schemes provide high data rates with a given transceiver complexity. Therefore, to some extent, MIMO architectures combined with UWB techniques can be a possible way to achieve very high data rates. Therefore, this makes possible a new range of short-distance, high-data-rate demanding applications, such as ubiquitous wire replacements or wireless PC connections (wireless USB).

A first step toward the integration of a MIMO architecture into a UWB communication system has been undertaken by the work reported in [133]. It proposed an IR system exploiting two transmit antennas and one receive antenna, employing a STBC scheme, over a simplistic channel model. This work exhibits an improvement of performance in terms of the BER. In another investigation we combined a multi-antenna STBC-IR system with orthogonal pulses, enhanced the data rate, introduced STTC for UWB, and reported the performance of both schemes over a realistic UWB channel model [134].

A further step has been achieved by comprehensive analysis of a STTC scheme in a UWB environment by adapting the scheme presented in [131] to single-band UWB signalling [135]. In this work a new performance criterion regarding the UWB environment and signalling properties has been derived. Generalization of mathematical derivations was also provided, and overall system performances were assessed. The theoretical analysis was supported by simulations undertaken for various situations. Eventually, results demonstrated the scale of achievable enhancement by a STTC-IR scheme against a single input single output (SISO)-IR and a STBC-IR scheme.

11.4 SELF-INTERFERENCE IN HIGH-DATA-RATE UWB COMMUNICATIONS

The UWB multipath channel may cause self-interference in high-data-rate UWB communications, which has generally been ignored in UWB studies under some simplifying assumptions.

Under the assumption that the delay resolution of the channel equals the duration T_w of the modulated pulses, multipath components would be at least T_w apart in arrival times and would not overlap at the receiver. However, it is not the duration T_w but the bandwidth B of the UWB pulses that decides the delay resolvability of a UWB channel, and the multipath delay resolution Δ_τ is approximately $1/B$ [50].

Actually, the fine delay resolution provided by the wide bandwidth of UWB signals may affect the performance of a correlator-based receiver. That is, if $\Delta_\tau < T_w$, pulses arriving at the receiver may overlap with each other and the pulse correlator output will be corrupted by correlations of the template waveform with interfering pulses that arrive or have tails within the correlator integration time, which is the duration of the desired pulse. The longer the pulse duration T_w, the more possible correlations of the template waveform with overlapping pulses. Note that almost all published UWB pulses to date, for example [66] and [69], have effective durations larger than the inverse of their bandwidths, i.e. $T_w > 1/B \approx \Delta_\tau$; therefore, these pulses will overlap when arriving at the receiver and lead to IPI.

Conventional rake structures are based on the path resolvability assumption that the minimum path spacing is larger than the signaling waveform's autocorrelation time, but path resolvability cannot be ensured in UWB wireless communications due to the extremely fine delay resolution of the UWB channel.

A rake structure with resolution reduction of the fingers has been proposed [136], which consists of fingers that are spaced less than one pulse duration apart. Due to the resolution reduction, the interference and noise components at the output of the rake will be correlated. In this case, the optimal combiner would be the minimum-mean-square-error combiner, for which the combining coefficients rely on the cross-correlation matrix of the interference and the noise components of the rake output. However, this correlation matrix is not available to the receiver in practice. A practical solution is the MRC rake receiver with resolution reduction.

To better understand the self-interference in high-rate UWB communications, consider a tapped delay line (TDL) channel model, for which the impulse response is given by

$$h(t) = \sum_{l=0}^{L-1} \alpha_l \delta(t - l\Delta_\tau) \tag{11.1}$$

where α_l are the path gain coefficients, $\Delta\tau$ is the delay resolution, with T_d denoting the channel delay spread, $L = \lceil T_d/\Delta\tau \rceil$ (in which $\lceil \cdot \rceil$ is the integer ceiling), and $\alpha_l = 0$ for any $l\Delta\tau$ that has no path arrivals. In this discrete channel model, all multipath components arrive at integer multiples of the delay resolution $\Delta\tau$.

If the pulse repetition time (or frame time) T_f is shorter than the channel delay spread T_d, channel responses to consecutively transmitted pulses will overlap, resulting in inter-frame interference (IFI). Such overlaps between channel responses are illustrated in Figure 11.5, in which the pulse repetition time T_f is equal to 10 ns while the channel delay spread is larger than 50 ns. At the same time, if the effective duration of the modulated pulses is larger than the delay resolution of the channel, i.e. $T_w > \Delta\tau$, pulses arriving at the receiver may overlap and lead to IPI.

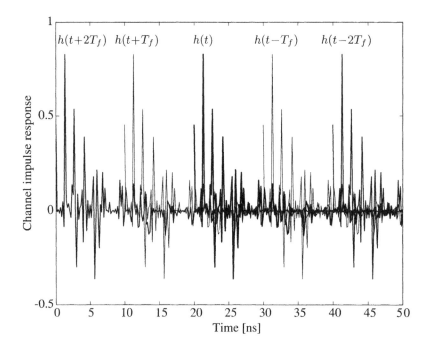

Fig. 11.5   Channel response overlaps that lead to IFI interference ($T_f = 10$ ns).

Figure 11.6 illustrates the overlaps between pulses arriving at the receiver, where the received pulse shape is assumed to be the Gaussian waveform $y_{g_3}(t)$ with a duration of $T_w = 0.5$ ns and the channel's delay resolution $\Delta\tau = 0.167$ ns corresponding to a signal bandwidth of $B = 6$ GHz.

High-order orthogonal modulations are preferable for high-data-rate communications. Consider $2M$-ary bi-orthogonal pulse modulation (BOPM), based on the use of M orthogonal pulses along with their negative versions. 4-ary BOPM can employ two Gaussian waveforms $y_{g_n}(t)$ and $y_{g_{n+1}}(t)$, since two successive derivatives of the generic Gaussian function are even (odd) and odd (even) functions, respectively, and thus are orthogonal to each other. 8-ary BOPM can use four orthogonal MHPs.

The two Gaussian waveforms $y_{g_5}(t)$ and $y_{g_6}(t)$ with $\tau = 0.0726$ ns have durations of around 0.5 ns and -10 dB bandwidths of about 7 GHz, corresponding to a delay resolution of 0.143 ns. The four MHPs $\{h_0(t), h_1(t), h_2(t), h_3(t)\}$ with $\tau = 0.1$ ns

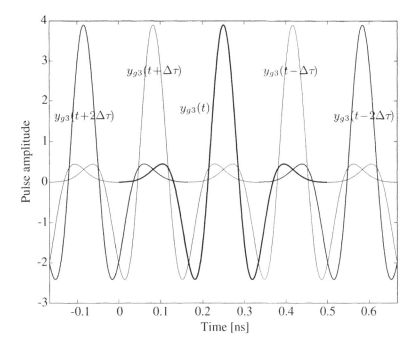

Fig. 11.6 Pulse overlaps that lead to IPI ($T_w = 0.5$ ns, $\Delta\tau = 0.167$ ns).

have durations of around 1 ns and a maximum -10 dB bandwidth of 7 GHz. These pulses are used to obtain the following numerical simulation results [137].

Figure 11.7 shows the BER of 4-ary BOPM, 8-ary BOPM, and bi-polar pulse amplitude modulation (BPAM) for a bit rate of 500 Mbps, under a channel model with a delay resolution of $\Delta\tau = 0.167$ ns [50]. The corresponding BER curves obtained by assuming that the multipath delay resolution $\Delta\tau$ equals the pulse duration T_w (dashed lines) are also plotted for comparison. The performance of each modulation under the realistic channel model is always degraded compared with its performance under the assumption of $\Delta\tau = T_w$ (i.e. no IPI). This reveals that the fine delay resolution provided by the wide bandwidth of UWB signals causes pulse overlap at the receiver and this IPI degrades the system performance.

Figure 11.8 shows the BER of 4-ary BOPM, 8-ary BOPM, and BPAM for the case with a pulse repetition time of $T_f = 2$ ns and one pulse per symbol. Under the assumption of no pulse overlap (dashed lines), 8-ary BOPM offers the best performance and achieves the highest data rate (1.5 Gbps) at the same time. However, since the multipath channel causes the four Hermite waveforms to overlap at the receiver, the actual performance (solid line) of 8-ary BOPM degrades catastrophically. For modulations using the two Gaussian waveforms, however, the performance degradation caused by pulse overlap is much less significant. This can be explained by the fact that orthogonal modified Hermite waveforms are nonzero for longer durations than Gaussian waveforms and their auto- and cross-correlation characteristics do not

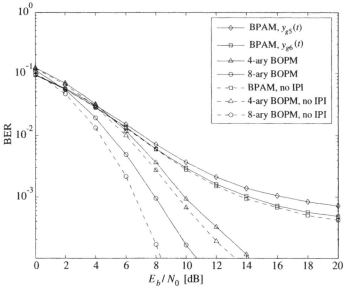

Fig. 11.7 The effect of IFI and IPI on BPAM, 4-ary BOPM, and 8-ary BOPM, for a data rate of 500 Mbps.

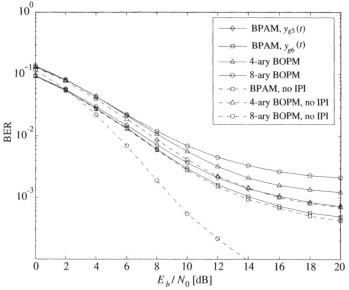

Fig. 11.8 The effect of IFI and IPI on BPAM, 4-ary BOPM, and 8-ary BOPM, for a pulse repetition time of 2 ns and one pulse per symbol.

appear as desirable as those of Gaussian waveforms. Therefore, to fully exploit the high data rate and power efficiency promised by high-level OPMs, the choice of signaling pulses is very important. In both Figures 11.7 and 11.8, BPAM using $y_{g_6}(t)$ performs better than BPAM using $y_{g_5}(t)$. This is because the autocorrelation of $y_{g_6}(t)$ has a narrower peak than that of $y_{g_5}(t)$. Hence, for modulations requiring only one pulse shape, for example BPAM, $y_{g_6}(t)$ is preferable to $y_{g_5}(t)$.

IFI cannot be avoided in high-data-rate UWB communications where high pulse repetition rate is required. The effect of pulse overlap can be made insignificant by using narrow pulses, but for high-order OPMs where longer-duration pulses cannot be avoided, the effect of IPI could be significant. Therefore, IFI and IPI may limit the performance of high-data-rate UWB communications.

11.5 COEXISTENCE OF DS-UWB WITH WI-MAX

One of the systems whose operation could potentially be harmed by UWB interference is the FBWA system which uses the IEEE 802.16 standard. This standard and its associated industry consortium, Wi-Max, aim to deliver high data rates over large geographical areas to a large number of users [138]. This means that Wi-Max will be looking to provide wireless last-mile broadband access with performance comparable to cable and DSL connections. It will provide connectivity with speeds up to 75 Mbps over ranges up to 30 miles, to underdeveloped areas as well as urban areas in which the broadband access cannot be provided due to problems such as excessive construction costs.

The first version of IEEE standard 802.16 was completed in October 2001. It defined the air interface and MAC protocol for a wireless metropolitan area network (WMAN), intended to provide high-bandwidth wireless connections for both residential and enterprise use. IEEE working group 802.16, which is responsible for the development of this standard, focused their initial interest on the 10–66 GHz range of the bandwidth. This band would only be suitable for LOS communication and is inapplicable for urban use. This shifted the attention and the activity of the group to the 2–11 GHz band, which led to the development of the IEEE 802.16a amendment that was completed in January 2001. Finally, the latest upgrade to IEEE 802.16a was published in January 2004. This upgrade is referred to as IEEE 802.16-2004. The development of IEEE standard 802.16-2004 has made sure that these systems hit the consumer market in the near future.

Four modes of operation specified in the standard for this technology will operate in the 2–11 GHz portion of the spectrum. Future developments of this standard (IEEE 802.16e) will focus on supporting mobility up to speeds of 70–80 mi/h, which will enable the subscriber station to have broadband wireless access on their PDA or laptops.

The operation of UWB systems has been restricted in the 3.1–10.6 GHz frequency range by the FCC. This means that they will share the bands used by IEEE-802.16 systems. The unlicensed use of this portion of the spectrum by UWB devices may pose a threat to FBWA systems based on Wi-Max. Therefore, if UWB technology is to be successful, it is vital for UWB developers to ensure that these devices pose no threat to the operation of FBWA receivers.

This section aims to investigate the impact of DS-UWB systems deployed in a realistic hot spot scenario on the operation of FBWA systems that are based on the IEEE 802.16 standard. This task will be accomplished through the following steps:

1. calculating the maximum allowable interference ($I_{\max}$) for each mode of operation set in IEEE-802.16;

2. developing an analytical expression for the interference produced by UWB devices;

3. defining a realistic hot spot scenario for UWB system;

4. calculating the interference caused by UWB devices for different factors;

5. comparing these values with the threshold values calculated in the first step.

In this section interference thresholds will be calculated using the IEEE 802.16 specifications, the signal model for DS-UWB will be defined, an analytical interference model will be developed, the hot spot scenario will be discussed, and finally UWB interference results acquired from simulations will be presented.

11.5.1 Interference thresholds

In this subsection the maximum tolerable interference on a system using IEEE 802.16 will be evaluated. In the absence of interference, the only factor contributing to loss in the received signal power is the thermal noise, which depends on the receiver characteristics and can be calculated using

$$N = -144 + 10\log(\text{BW}_{\text{RX}}) + N_F \tag{11.2}$$

in which BW_{RX} is the receiver bandwidth and N_F is its noise figure. In the presence of interference due to UWB systems I_{UWB}, the received SNR, will be reduced further. If we consider SNR to be the signal-to-noise ratio in the absence of interference and SINR the signal-to-interference plus noise ratio,

$$\text{SINR} = \frac{S}{I_{\text{UWB}} + N} \tag{11.3}$$

then the reduction in SNR due to UWB interference could be referred to as degradation d and calculated in the following manner:

$$d = \frac{\text{SNR}}{\text{SINR}} = \frac{I_{\text{UWB}} + N}{N} \tag{11.4}$$

This reduction will worsen the performance of the system. The aim of this sub-section is to calculate the maximum I_{UWB} in the presence of which the performance of the victim receiver (IEEE 802.16 receiver) is still acceptable.

Several modes of operation have been specified for Wi-Max. The first mode which will be used for LOS communications operates in frequency bands above 11 GHz.

As the bands predicted for this mode do not overlap with the bands specified by the FCC for UWB, its operation will not be harmed by the UWB interference.

Of the next five modes, two (WirelessMAN-SCa and WirelessMAN-OFDM) will be discussed here. The reason for this is that the operation of these two modes have a higher potential of being distorted by UWB interference. The interference threshold for these two modes will be evaluated in the following sub-sections.

11.5.1.1 WirelessMAN-SCa This mode is based on single carrier technology for NLOS operation in the sub-11 GHz bands and will operate in licensed bands. The minimum allowable bandwidth here is 1.25 MHz and a 1 dB degradation in SNR due to interference is acceptable for this mode. In this way, I_{max} can be calculated using Equation (11.3). The results have been reflected in Table 11.1. It should be mentioned that the receiver noise figure has been considered to be 7 dB.

Table 11.1 Maximum tolerable interference in the WirelessMAN-SCa mode for different bandwidths.

	BW [MHz]				
	1.25	2.5	5	10	20
N [dBW]	−136	−133	−130	−127	−124
I_{max} [dBW]	−141.9	−138.8	−135.9	−132.9	−129.9

11.5.1.2 WirelessMAN-OFDM This mode was the mandatory mode in the IEEE 802.16a amendment. Like WirelessMAN-SCa, it is designed for operation in the sub-11 GHz band and uses OFDM technology for NLOS communication scenarios. It has been specified that for satisfactory performance the BER should be less than 10^{-6} at given SNR values and modulation code rates. Concatenated Reed–Solomon convolutional coding is used for data transmission in this mode, and minimum E_b/N_0 in order to have BER $< 10^{-6}$ is determined from BER curves in [139]. According to this work, in order for BER to be 10^{-6}:

$$\left(\frac{E_b}{N_0}\right)_{\text{rate}=1/2} = 3.375 \text{ dB} \tag{11.5}$$

$$\left(\frac{E_b}{N_0}\right)_{\text{rate}=2/3} = 3.750 \text{ dB} \tag{11.6}$$

$$\left(\frac{E_b}{N_0}\right)_{\text{rate}=3/4} = 4.125 \text{ dB} \tag{11.7}$$

The SNR assumptions would yield BER levels far less than the levels specified in the above equations. This would provide us with an interference margin. This margin I_{max}

is a function of thermal noise at the receiver, N_0, the minimum allowed bit energy at receiver input, $E_{b,\min}$, and the satisfactory bit energy at the receiver input for which the BER is less than 10^{-6}, $E_{b,\text{sat}}$:

$$I_{\max} = N_0 \left(\frac{E_{b,\min}}{E_{b,\text{sat}}} - 1 \right) \tag{11.8}$$

N_0 can be calculated using Equation (11.2). Given the specifications for satisfactory performance in [139], both $E_{b,\min}$ and $E_{b,\text{sat}}$ will be calculated as well. Having these values, $I_{\max}$ will be known for different channelization and coding rates. The results for $I_{\max}$ have been reflected in Table 11.2.

Table 11.2 Maximum tolerable interference in the WirelessMAN-OFDM mode for different channelization and coding rates 1/2 and 3/4.

	BW [MHz]				
	1.75	3.5	7	10	20
N [dBW]	−134.5	−131.5	−128.5	−127.0	−123.9
$I_{\max}$ [dBW], rate 1/2	−126.9	−123.9	−120.9	−121.18	−118.2
$I_{\max}$ [dBW], rate 3/4	−129.6	−126.6	−123.6	−122.1	−119.1

11.5.2   UWB signal model

The UWB system used in this analysis is DS-UWB. The baseband signal in this model can be written as

$$S_{\text{UWB}} = \sum_{k=-\infty}^{\infty} c_k p(t - kT_c) \tag{11.9}$$

where $p(t)$ is the baseband pulse transmitted in each chip interval of duration T_c and c_k is the amplitude of the transmitted chip. Here, two different pulse shapes have been considered for $p(t)$. These two pulse shapes are a Gaussian pulse and a Gaussian monocycle as shown in Equations (2.5) and (2.6), respectively:

$$p_1(t) = K_1 e^{-(t/\tau)^2} \tag{11.10}$$

$$p_2(t) = K_2 \frac{-2t}{\tau^2} e^{-(t/\tau)^2} \tag{11.11}$$

where K_1 and K_2 are constants and can be easily calculated using the permitted energy (specified by the FCC spectral mask) for the pulse shape. The parameter τ is the scaling factor.

Assuming that c_k is an ideal random bipolar zero-mean white sequence with variance σ, then the power spectrum of the baseband UWB signal will be

$$S(f) = \frac{\sigma^2}{T_c}|P(f)|^2 \tag{11.12}$$

where $P(f)$ is the Fourier transform of the transmitted pulse shape and

$$P_1(f) = K_1 \tau \sqrt{\pi} e^{-(\pi \tau f)^2} \tag{11.13}$$

$$P_2(f) = K_2 \tau \sqrt{\pi} (j2\pi f) e^{-(\pi \tau f)^2} \tag{11.14}$$

To comply with the FCC bandwidth regulations, $P(f)$ should be shifted up to the allowed frequency band with a center frequency f_c. Therefore, the power spectrum of the UWB signal will be

$$S_{\text{UWB}}(f) = \tfrac{1}{2}\{S(f - f_c) + S(f + f_c)\} \tag{11.15}$$

As the victim system is narrowband, only a portion of the UWB power that lies in the victim bandwidth would be received by the system. If we neglect the propagation loss for the moment, the interference power received at the victim receiver due to the ith UWB device can be expressed as

$$I_{\text{UWB},i} = \int_{f_{\text{low}}}^{f_{\text{high}}} S_{\text{UWB}}(f)\, df \tag{11.16}$$

where f_{low} and f_{high} are the lower and the upper frequency bounds between which the victim receiver operates.

11.5.3 Interference model

A mathematical expression for the UWB received power at the terminals of the victim receiver can now be developed. The UWB device signal power S_{RX} at the victim receiver can be expressed as

$$S_{\text{RX}}(f) = \frac{A_f G_{\text{TX}} G_{\text{RX}} S_{\text{UWB}}(f)}{P_L(r)} \tag{11.17}$$

where G_{TX} and G_{RX} are the transmitter (UWB device) and the receiver (Wi-Max system) antenna gains. It is assumed that both antennas are omnidirectional with unity gains. A_f is the activity factor and represents the percentage of active UWB devices at an instant. $P_L(r)$ is the path loss which depends on the distance r between the UWB transmitter and the victim receiver. A free space propagation model has been used here for path loss calculations. Clearly, UWB devices will operate indoors and, as the victim receiver is mounted outside the building, the penetration loss due to the walls should also be considered in the model [140]:

$$P_L(r) = a_0 L_p r^2 \tag{11.18}$$

where $a_0 = (4\pi/\lambda)^2$ and L_p represents the penetration loss which has been considered to be 10 dB. Consider n UWB devices distributed between $[r_{\min}, r_{\max}]$ in the vicinity of the victim receiver. The average transmit power density of UWB devices can be written as

$$\rho_I = \frac{\sum_{i=1}^{n} S_{\text{TX},i}}{A} \tag{11.19}$$

where A represents the area of the room in which the hot spot is located and $S_{\text{TX},i}$ is the transmit power of the ith UWB device. Therefore, the interference experienced by the victim receiver can be expressed as

$$I = \int_{f_{\text{low}}}^{f_{\text{high}}} \int_S \frac{A_f G_{\text{TX}} G_{\text{RX}} \rho_I}{a_0 L_p r^2} \, ds \, df \tag{11.20}$$

Therefore by using $ds = r \, dr \, d\varphi \, df$,

$$I = \int_{f_{\text{low}}}^{f_{\text{high}}} \int_{\varphi=0}^{2\pi} \int_{r=r_{\min}}^{r_{\max}} \frac{A_f G_{\text{TX}} G_{\text{RX}} \rho_I}{a_0 L_p r^2} \, r \, dr \, d\varphi \, df \tag{11.21}$$

which would lead to

$$I = K \ln \left(\frac{r_{\max}}{r_{\min}} \right) \int_{f_{\text{low}}}^{f_{\text{high}}} S_{\text{UWB}}(f) \, V \, df \tag{11.22}$$

where $K = (2n A_f G_{\text{TX}} G_{\text{RX}})/(a_0 L_p R^2)$, R is the radius of the area containing the UWB devices, and $r_{\max}$ and $r_{\min}$ are the maximum and minimum distances that the UWB devices can be from the victim receiver.

After calculating the portion of S_{UWB} that lies in the bandwidth of the victim receiver, the total interference power at the victim receiver for a Gaussian pulse can be written as

$$I_{1,\text{UWB}} = C_1 \ln \left(\frac{r_{\max}}{r_{\min}} \right) \left[\sqrt{\frac{\pi}{2}} \, \text{erf} \left(\tau \sqrt{2} \pi (f - f_c) \right) \right]_{f_{\text{low}}}^{f_{\text{high}}} \tag{11.23}$$

where $C_1 = (\pi \tau \sigma_c^2 K_1^2 K)/(4 T_c)$. In the same way, the total interference power at the victim receiver for a Gaussian monocycle can be written as

$$I_{2,\text{UWB}} = C_2 V \ln \left(\frac{r_{\max}}{r_{\min}} \right) \tag{11.24}$$

where

$$V = \left[\sqrt{2\pi} \, \text{erf}(\sqrt{2}\pi\tau(f - f_c)) - 4\pi\tau(f - f_c) e^{-2\pi^2 \tau^2 (f - f_c)^2} \right]_{f_{\text{low}}}^{f_{\text{high}}} \tag{11.25}$$

and $C_2 = (\pi \sigma_c^2 K_2^2 K)/(4 \tau T_c)$.

11.5.4   Interference scenario

The analysis have been based on the results produced by a number of simulation programs along with the analytical results produced by Equations (11.23) and (11.24). Each time that the simulation programs are executed, a specified number of UWB devices will be randomly located in the hot spot. This means that the simulator will generate a unique realistic hot spot scenario upon execution. A 10 m by 10 m room is the physical location in which the hot spot forms. The FBWA receiver which could be operating in any of the modes described has been considered to be mounted on top of this room.

In order to calculate the total interference resulting from the UWB devices, the interference resulting from each device must be calculated. The aggregate interference at the victim receiver can be computed as the summation of these individual interference values:

$$I_{\mathrm{UWB}} = \sum_{i=1}^{n} \int_{f_{\mathrm{low}}}^{f_{\mathrm{high}}} S_{\mathrm{RX}} \, f \, df \qquad (11.26)$$

The criteria used to decide whether I_{UWB} would harm the operation of a Wi-Max system or not are the I_{max} values calculated in Tables 11.1 and 11.2. As long as the interference resulting from the UWB hot spot is maintained below these values, it can be concluded that the UWB hot spot poses no threat to the operation of the Wi-Max receiver.

11.5.5   Some numerical results

In this section the data resulting from simulation of various scenarios will be analyzed. Results have been generated by varying different parameters including victim receiver bandwidth, carrier frequency, activity factor, and number of UWB devices. The results will be compared with the I_{max} values in Tables 11.1 and 11.2 in order to show the circumstances under which the interference produced in these scenarios will be harmful to the victim receiver which is a Wi-Max/IEEE-802.16-based FBWA receiver.

As the receiver bandwidth of a Wi-Max device depends on regulatory laws, we can have different receiver bandwidths according to these regulations. Therefore, it is vital to study the effect of receiver bandwidth on the produced interference. The results of such an analysis have been shown in Figure 11.9.

The values of I_{max} in Tables 11.1 and 11.2, corresponding to WirelessMAN-SCa and WirelessMAN-OFDM, have also been included for comparison in Figure 11.9. The figure shows that the aggregate interference increases with bandwidth. As can be seen, in this scenario (50 UWB devices with 50% activity factor), the UWB hot spot poses no threat to the operation of any of these modes and the aggregate interference falls below the threshold values for both the Gaussian pulse and Gaussian monocycle.

It should be noted that, in practical cases, having 50 UWB devices with 50% activity factor in a 10 m by 10 m room is a very pessimistic assumption. In reality, lower values are expected for both the number of devices and activity factor. The results for the two different carrier frequencies in this figure show that for lower f_c

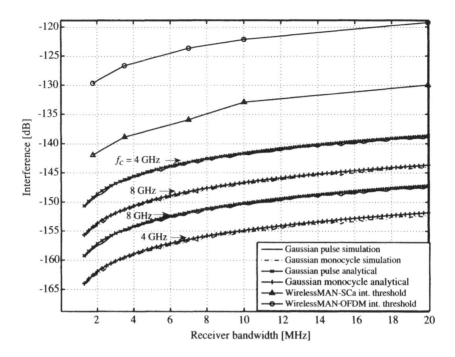

Fig. 11.9 Comparison between aggregate interference with threshold values for WirelessMAN-SCa and WirelessMAN-OFDM for two values of f_c for both pulse shapes. $n = 50$, $A_f = 50\%$.

values the Gaussian monocycle has a better performance while for higher f_c values the Gaussian pulse produces less interference.

Figure 11.10 shows results for the interference generated by the UWB hot spot for different numbers of UWB devices. In this step, the aggregate interference of a number of randomly positioned UWB devices will be calculated. The figure shows that as the number of UWB devices increases the probability of having harmful interference becomes higher.

As can be seen, for the case of the Gaussian pulse with 50% activity factor, $I_{\max}$ reaches the threshold values in Table 11.1 (WirelessMAN-SCa) as the number of users approaches 50. This means that to avoid harmful interference for this case the number of UWB devices should be kept below a minimum. This minimum value is directly dependent on the layout of UWB devices inside the room or, in other words, on their average distance from the victim receiver.

11.5.6 Conclusion

In summary, in a realistic hot spot scenario (with a reasonable number of UWB devices that are active in a reasonable period of time, distributed uniformly on the floor), UWB devices pose no threat to the operation of IEEE 802.16 systems and the

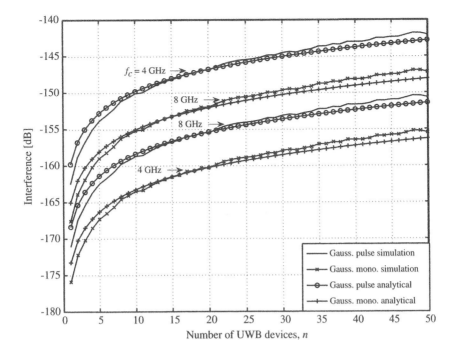

Fig. 11.10 Aggregate interference versus the number of users for both the Gaussian pulse and monocycle for two values of f_c. $A_f = 50\%$, BW=10 MHz.

aggregate interference caused by them will fall below the thresholds specified in the standard.

It should be noted, however, that increasing the number of UWB devices or placing the majority of devices close to the victim receiver might damage the operation of the receiver. It can also be concluded that having high activity factors would increase the chances of having harmful interference. These problems could be tackled by developing some form of power control for the UWB system.

It can be concluded that, of the two IEEE 802.16 modes studied here, the mode that might have problems in the vicinity of a UWB hot spot is WirelessMAN-SCa when operating in low bandwidths, as it has a threshold much lower than the other proposed modes. The analysis showed, however, that in realistic and practical cases a Wi-Max/IEEE 802.16-based system can operate in all modes without any problems in the presence of a UWB hot spot.

11.6 VEHICULAR RADARS IN THE 22–29 GHZ BAND

Vehicular radar in the allocated 22–29 GHz band is among the technologies considered to permit the operation and marketing of several new devices using UWB technology.

In the past, the FCC has considered the implementation of vehicular radar at 47, 60, and 76 GHz. Vehicular radar is intended for collision avoidance, improved airbag functioning, and improved suspension systems. An implementation may include many transmitters per vehicle mounted on bumpers and fenders. Millimeter wavelengths are preferred for operation because of the need for a small antenna and small beam, and because of the need for significant attenuation at increasing distances. The emitted signal is pulsed in short bursts.

Employment of vehicular radar devices operating around 24 GHz has been promoted for deployment since 2003 onwards. These sorts of short-range radars combine two features:

- high-resolution distance measurement providing speed information of an approaching mobile object using Doppler radar based on a narrowband signal falling within the short-range device band 24.05–24.25 GHz;

- a wide band radar providing the position of objects to a resolution of a couple of centimeters using a bandwidth of approximately 5 GHz around 24 GHz.

Vehicular radar systems are limited to operation with a center frequency greater than 24.075 GHz. Furthermore, it is required that the frequency at which the highest radiated emission level occurs must also be greater than 24.075 GHz and that the -10 dB bandwidth be contained between 22 GHz and 29 GHz. This is high enough in frequency to ensure antenna directionality along with a high level of signal attenuation with increasing distance and intervening objects. It is also high enough in frequency to permit the use of an antenna small enough to be mounted on an automobile. Table 11.3 shows the FCC spectral mask for UWB vehicular radar applications.

Table 11.3 FCC spectral mask for UWB vehicular radar.

Frequency band [GHz]	0.96–1.61	1.61–22	22–29	29–31	>31
EIRP [dBm/MHz]	−75.3	−61.3	−41.3	−51.3	−61.3

11.6.1 Environment sensing for vehicular radar

It is a well-known estimate that more than 90% of all road accidents involve some human error and in most accidents a human is solely to blame. The idea of introducing a variety of sensing technologies to provide a vehicle with a full environmental awareness as a means of improving safety arises from these interesting statistics. Almost all collisions could be avoided if driver reaction time was shifted forward by 2 s. It is predicted that almost all of these accidents could be avoided through the introduction of automatic vehicle intervention through the use of a hybrid sensor array to form a safety belt around the vehicle. A typical schematic of one of these sensor arrays is shown in Figure 11.11.

The main requirement of these applications is the range resolution. The ability to precisely determine an object's location is inversely proportional to the signal bandwidth. The precise measurement of object movement is essential for the prediction of trajectories and the prevention of false alarms. For a maximum resolution of 10 cm the spectral occupancy of the main lobe of the incident wave has to be at least 3 to 4 GHz wide to allow for errors, variations, and design margin.

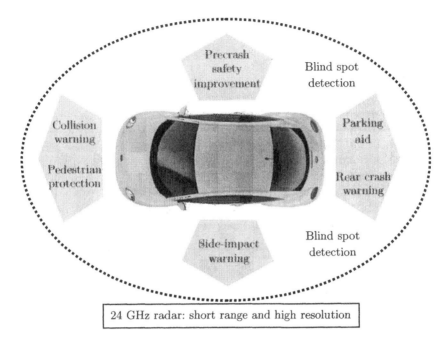

Fig. 11.11 Environment sensing for automobiles is achieved through an array of multiple sensors.

The availability of such a broad spectral frequency band has to be balanced with other practical considerations. The gain or focusing ability of an antenna is proportional to the antenna aperture. The antenna of each sensor, therefore, has to be small enough that an aperture of suitable electrical size can be physically mounted in the restrictive space behind a vehicle's bumper fascia. In addition, as a percentile of center frequency, it becomes easier to produce antennas that support signals at least 4 GHz in width by increasing the frequency of operation. However, there are also several practical considerations that conversely argue for as low an operating frequency as possible [141].

There are several licensed frequency bands that surround the 24.125 GHz band that are used for low-power applications. In particular, a global spectrum allocation between 23.6 GHz and 24 GHz is used for radio astronomy and remote sensing. It is this band, in particular, that has caused consternation amongst other spectrum users, particularly the European Space Agency (ESA), the European Meteorological

Satellite Service (EUMETSAT), the Committee on Radio Astronomy Frequencies (CRAF), and others.

There are also concerns about the effect of increases in wideband spectral emissions on fixed service communications (also in the 23–26 GHz band in Europe). This has led to the specific regulations imposed by the FCC, which calls for reduced emissions at certain elevation angles above the horizon over time. The specification is written in terms of an emitted PSD, or EIRP, which allows for either a reduction in the peak transmitted power or, as is more likely, a reduction in the elevation sidelobes of the transmitter antenna.

11.7 SUMMARY

In this chapter we considered a number of new topics that have been under investigation for UWB communication systems over the last few years. First, we started with the application of UWB techniques in ad-hoc and sensor networks within a small range inside a room or building. Second, space-time coding for pulse-based UWB systems was explained. Then, self-interference in high-data-rate UWB communications was discussed. As a possible victim communication system subject to the interference from UWB signals, Wi-Max interference issues were considered in more detail. Finally, vehicular radar in the 22–29 GHz band and their applications were discussed.

Problems

Problem 1. With reference to Figure 11.3 explain the code sense technique proposed as the optimal MAC protocol for DS-UWB systems.

Problem 2. What is a body area network (BAN) and how it can be helpful in hospitals?

Problem 3. What are the advantages of space-time coding for UWB systems?

Problem 4. How is self-interference defined in high-rate UWB communications?

Problem 5. How many modes are defined for Wi-Max? Briefly explain two of them.

Problem 6. What are the advantages of the 22–29 GHz bandwidth of vehicular radar systems?

References

1. J. D. Taylor, editor. *Ultra-Wideband Radar Technology*. CRC Press, 2001.

2. J. D. Taylor, editor. *Introduction to Ultra-Wideband Radar Systems*. CRC Press, 1995.

3. G. F. Ross. Transmission and reception system for generating and receiving base-band duration pulse signals without distortion for short base-band pulse communicaton system. *US Patent 3,728,632*, April 1973.

4. Time Domain. http://www.timedomain.com.

5. XtremeSpectrum. http://www.xtremespectrum.com.

6. T. W. Barrett. History of ultra wideband (UWB) radar & communications: Pioneers and innovators. In *Proceedings of Progress in Electromagnetics Symposium 2000 (PIERS2000)*, July 2000.

7. C. L. Bennett and G. F. Ross. Time-domain electromagnetics and its applications. *Proceedings of the IEEE*, 66:299–318, March 1978.

8. J. Williams. The IEEE 802.11b security problem, part 1. *IT Professional*, pages 91–96, November 2001.

9. F. Ramirez-Mireles and R. A. Scholtz. Wireless multiple-access using SS time-hopping and block waveform pulse position modulation, part 2: Multiple-access performance. In *Proceedings ISITA Symposium*, October 1998.

Ultra Wideband Signals and Systems in Communication Engineering   Second Edition
M. Ghavami, L. B. Michael and R. Kohno   © 2007 John Wiley & Sons, Ltd

10. M. Z. Win and R. A. Scholtz. Ultra-wide bandwidth time-hopping spread spectrum impulse radio for wireless multiple-access communication. *IEEE Transactions on Communications*, 48(4):679–691, April 2000.

11. D. G. Leeper. Wireless data blaster. *Scientific American*, May 2002.

12. H. Kikuchi. UWB arrives in Japan. *Nikkei Electronics*, pages 95–122, February 2003.

13. R. Mark. XtremeSpectrum rolls out first UWB chipset. InternetNews Website, June 2002.

14. *FCC regulations, 47CFR Section 15.5 (d)*. http://ftp.fcc.gov, 1998.

15. R. A. Scholtz. Multiple access with time-hopping impulse modulation. In *IEEE MILCOM93, vol. 2*, October 1993.

16. J. T. Conroy, J. L. LoCicero, and D. R. Ucci. Communication techniques using monopulse waveforms. In *IEEE MILCOM99, vol. 2*, November 1999.

17. M. Ghavami, L. B. Michael, S. Haruyama, and R. Kohno. A novel UWB pulse shape modulation system. *Kluwer Wireless Personal Communications Journal*, 23:105–120, 2002.

18. J. Mills, editor. *Radio Communication Theory and Methods*. McGraw-Hill, 1917.

19. J. M. Wilson. Ultra wideband technology update at spring 2003. *Intel Developer UPDATE Magazine*, pages 1–9, 2003.

20. New ultra-wideband technology, white paper. *Discrete Time Communications*, pages 1–8, 2002.

21. H. F. Harmuth. Radio signals with large relative bandwidth for over-the-horizon radar and spread spectrum communications. *IEEE Transactions on Electromagnetic Compatibility*, 20:501–512, 1978.

22. J. R. Davis, D. J. Baker, J. P. Shelton, and W. S. Ament. Some physical constraints on the use of carrier free waveforms in the radio-wave transmission systems. *Proceedings of IEEE*, 67:884–890, June 1979.

23. H. P. Hsu. *Analog and Digital Communications*. McGraw-Hill, 1993.

24. A. V. Oppenheim, A. S. Willsky, and I. T. Young. *Signals and Systems*. Prentice Hall, 1983.

25. A. V. Oppenheim and R. W. Schafer. *Discrete-Time Signal Processing*. Prentice Hall, 1989.

26. P. P. Newaskar, R. Blazquez, and A. P. Chandrakasan. A/D precision requirements for an ultra-wideband radio receiver. In *SIPS 02*, October 2002.

27. W. Ellersick, C. K. Ken Yang, W. Horowitz, and W. Dally. Gad: A 12gs/s CMOS 4-bit A/D converter for an equalized multi-level link. In *Symposium on VLSI Circuits, Digest of Technical Papers*, 1999.

28. M. I. Skolnik. *Introduction to Radar Systems*. McGraw-Hill, 1962.

29. D. G. Fink and D. Christiansen. *Electronics Engineers Handbook*. McGraw-Hill, 1975.

30. T. E. McEwan. Ultra-wideband radar motion sensor. *US Patent 5,361,070*, 1994.

31. J. R. Foerster. The effects of multipath interference on the performance of UWB systems in an indoor wireless channel. In *Spring Vehicular Technology Conference*, May 2001.

32. H. Hashemi. Impulse response modeling of indoor radio propagation channels. *IEEE Journal on Selected Areas in Communications*, 11:967–978, 1993.

33. K. Pahlavan and A. Levesque. *Wireless Information Networks*. John Wiley & Sons, Inc., 1995.

34. M. Z. Win and R. A. Scholtz. On the robustness of ultra-wide bandwidth signals in dense multipath environments. *IEEE Communications Letters*, 2:10–12, 1998.

35. A. A. Saleh and R. A. Valenzuela. A statistical model for indoor multipath propagation. *IEEE Journal of Selected Areas in Communications*, 5:128–137, 1987.

36. H. Suzuki. A statistical model for urban radio propagation. *IEEE Transactions on Communications*, 25:673–680, 1977.

37. R. Ganesh and K. Pahlavan. Statistical modeling and computer simulation of indoor radio channel. *IEE Proceedings, part I(3)*, 138:153–161, 1991.

38. J. M. Cramer, R. A. Scholtz, and M. Z. Win. On the analysis of UWB communication channels. In *IEEE MILCOM99*, November 1999.

39. S. S. Ghassemzadeh, R. Jana, C. Rice, W. Turin, and V. Tarokh. A statistical path loss model for in-home UWB channels. In *IEEE UWBST*, May 2002.

40. J. Foerster and Q. Li. UWB channel modeling contribution from Intel. Technical report, IEEE document, 2002.

41. K. Siwiak and A. Petroff. A path link model for ultra wide band pulse transmission. In *IEEE Vehicular Technology Conference 2001*, pages 1173–1175, May 2001.

42. D. Cassioli, M. Z. Win, and A. R. Molisch. The ultra-wide bandwidth indoor channel: From statistical model to simulations. *IEEE Journal on Selected Areas in Communications*, 20:1247–1257, 2002.

43. A. Armogida, B. Allen, M. Ghavami, M. Porretta, and H. Aghvami. Path loss modeling in short-range UWB transmissions. In *International Workshop on UWB Systems, IWUWBS2003*, June 2003.

44. W. C. Stone. Nist construction automation report no. 3: Electromagnetic signal attenuation in construction materials. Technical report, BFRL Publications, 1997.

45. T. S. Rappaport. *Wireless Communications: Principles and Practice.* Prentice Hall, 1996.

46. W. Turin, R. Jana, S. S. Ghassemzadeh, C. W. Rice, and V. Tarokh. Autoregressive modeling of an indoor UWB channel. In *IEEE UWBST*, May 2002.

47. S. Howard and K. Pahlavan. Autoregressive modeling of wide-band indoor radio propagation. *IEEE Transactions on Communications*, pages 1540–1552, September 1992.

48. S. L. Marple. *Digital Spectral Analysis.* Prentice Hall, 1987.

49. M. Z. Win, R. A. Scholtz, and M. A. Barnes. Ultra-wide bandwidth signal propagation for indoor wireless communications. In *IEEE ICC 1997*, June 1997.

50. J. Foerster *et al.* Channel modeling sub-committee report (final). Technical report, IEEE P802.15 Wireless Personal Area Networks, February 2003.

51. A. F. Molisch *et al.* 802.15.4a channel model (final). Technical report, IEEE P802.15 Wireless Personal Area Networks, December 2004.

52. F. Héliot. *Design and Analysis of Space-time Block and Trellis Coding Schemes for Single-Band UWB Communications Systems.* PhD thesis, 2006.

53. J. G. Proakis. *Digital Communications.* Addison Wesley, 4th edition, 2000.

54. J. McCorkle. Why such uproar over ultrawideband? Communication Systems Design Website, March 2002. http://www.commsdesign.com/csdmag /sections/feature_article/OEG20020301S0021.

55. L. Zhao and A. M. Haimovich. The capacity of a UWB multiple access communication system. In *IEEE International Conference on Communications, ICC '02*, pages 1964–1968, May 2002.

56. K. Eshima, K. Mizutani, R. Kohno, Y. Hase, S. Oomori, and F. Takahashi. Comparison of ultra-wideband (UWB) impulse radio with DS-CDMA and FH-CDMA. In *Proceedings of 24th Symposium on Information Theory and Applications (SITA), Kobe, Japan*, pages 803–806, 2001. In Japanese.

57. T. Ikegami and K. Ohno. Interference mitigation study for UWB impulse radio. In *IEEE PIMRC 2003*, September 2003.

58. M. Hämäläinen, J. Saloranta, J. P. Makela, I. Opperman, and T. Pantana. Ultra wideband signal impact on IEEE 802.11b and bluetooth performance. In *IEEE PIMRC 2003*, September 2003.

59. M. Luo, M. Koenig, D. Akos, S. Pullen, and P. Enge. Potential interference to GPS from UWB transmitters phase II test results accuracy, loss-of-lock, and acquisition testing for GPS receivers in the presence of UWB signals. Technical report 3.0, Stanford University, March 2001.

60. J. P. Van't Hof and D. D. Stancil. Ultra-wideband high data rate short range wireless links. In *IEEE Vehicular Technology Conference*, pages 85–89, 2002.

61. K. Sarfaraz, A. Ghorashi, M. Ghavami, and H. Aghvami. Performance of Wi-Max receiver in presence of DS-UWB system. *IEE Electronics Letters*, 41:1388–1390, December 2005.

62. X. Chu and R. D. Murch. The effect of NBI on UWB time-hopping systems. *IEEE Transactions on Wireless Communications*, 3(5):1431–1436, September 2004.

63. E. Kreyszic. *Advanced Engineering Mathematics*. John Wiley & Sons Inc., 1988.

64. J. B. Martens. The hermite transform – theory. *IEEE Transactions on Acoustics, Speech and Signal Processing*, 38:1595–1606, 1990.

65. M. R. Walton and H. E. Hanrahan. Hermite wavelets for multicarrier data transmission. In *South African Symposium on Communications and Signal Processing ComSIG 93*, August 1993.

66. M. Ghavami, L. B. Michael, and R. Kohno. Hermite function-based orthogonal pulses for ultra wideband communication. In *WPMC'01*, September 2001.

67. M. Z. Win and R. A. Scholtz. Impulse radio: How it works. *IEEE Communications Letters*, 2:36–38, 1998.

68. D. Slepian. Prolate spheroidal wave functions, fourier analysis and uncertainty V: The discrete case. *Bell System Technical Journal*, 57, 1978.

69. R. S. Dilmaghani, M. Ghavami, B. Allen, and H. Aghvami. Novel pulse shaping using prolate spheroidal wave functions for UWB. In *IEEE PIMRC 2003 Beijing, China*, 2003.

70. C. Flammer. *Spheroidal Wave Functions*. Stanford University Press, 1957.

71. G. Arfken. *Mathematical Methods for Physicists*, chapter 12: Legendre Functions. Academic Press, 3rd edition, 1985.

72. N. W. Bailey. On the product of two Legendre polynomials. *Proceedings of the Cambridge Philosophical Society*, 29:173–177, 1933.

73. E. Hernandez and G. L. Weiss. *A First Course on Wavelets.* CRC Press, 1996.

74. C. K. Chui. *An Introduction to Wavelets.* Academic Press, 1992.

75. S. Ciolino, M. Ghavami, and H. Aghvami. On the use of wavelet packets in UWB pulse shape modulation systems. *IEICE Transactions on Fundamentals, special section on UWB*, 88(9):2310–2317, 2005.

76. D. K. Cheng. *Field and Wave Electromagnetics.* Addison-Wesley, 1989.

77. I. I. Immoreev and A. N. Sinyavin. Features of ultra-wideband signals' radiation. In *UWBST 2002 IEEE Conference on Ultra Wideband Systems and Technologies*, May 2002.

78. F. Sabath. Near field dispersion of impulse radiation. In *URSI General Assembly 2002*, August 2002.

79. C. Balanis. *Antenna Theory.* John Wiley & Sons Inc., 1997.

80. K. Y. Yazdandoost and R. Kohno. Ultra wideband antenna, CRL report.

81. Farr Research Inc. http://www.farr-research.com/.

82. S. Ramo and J. Whinnery. *Fields & Waves in Modern Radio.* John Wiley & Sons, Inc., 1962.

83. Fractus. http://www.fractus.com.

84. B. Widrow, P. E. Mantey, L. J. Griffiths, and B. B. Goode. Adaptive antenna systems. *Proceedings of IEEE*, 55:2143–2159, December 1967.

85. M. G. M. Hussain. An overview of the principle of ultra-wideband impulse radar. In *CIE 1996 International Conference of Radar*, November 1996.

86. M. G. M. Hussain. Antenna patterns of nonsinusoidal waves with the time variation of a gaussian pulse – part I. *IEEE Transactions on Electromagnetic Compatibility*, 30:504–512, 1988.

87. CDMA Development Group. CDG: Test plan document for location determination technologies evaluation, 2000.

88. N. Lenihan and S. McGrath. REALM: Analysis of alternatives for location positioning.

89. K. Pahlavan, X. Li, and J. Makela. Indoor geolocation science and technology. *IEEE Communications Society Magazine*, February 2002.

90. M. O. Sunay and I. Tekin. Mobile location tracking for IS-95 using the forward link time difference of arrival techniques and its application to zone-based billing. In *IEEE GLOBECOM Conference 1999*.

91. X. Li. *Super-Resolution TOA Estimation with Diversity Techniques for Indoor Geolocation Applications.* PhD thesis, 2003.

92. R. J. Fontana. Experimental results from an ultra wideband precision geolocation system. *Ultra-Wideband, Short-Pulse Electromagnetics IV.* Kluwer Academic/Plenum Publishers, May 2000.

93. R. J. Fontana and S. Gunderson. Ultra-wideband precision asset location system. In *UWBST 2002 IEEE Conference on Ultra Wideband Systems and Technologies*, May 2002.

94. I. Maravic, M. Vetterli, and K. Ramchandran. Channel estimation and synchronization with sub-Nyquist sampling and application to ultra-wideband systems. In *ISCAS 2004.*

95. R. Fleming and C. Kushner. Low-power miniature distributed position location and communication devices using ultra wideband, nonsinusoidal communication technology. Technical report, AetherwireLocation Inc., July 1995.

96. D. Porcino and W. Hirt. Ultra-wideband radio technology: Potential and challenges ahead. *IEEE Communications Magazine*, July 2003.

97. Intel Corporation. http://www.intel.com/technology/ultrawideband/.

98. Motorola Corporation. http://www.motorola.com/.

99. Communication Research Laboratory. http://www2.crl.go.jp/.

100. General Atomics. http://www.fusion.gat.com/photonics/uwb/.

101. Wisair. http://www.wisair.com.

102. Artimi. http://artimi.com.

103. Ubisense. http://www.ubisense.net.

104. M. Nakagawa, H. Zhang, and H. Sato. Ubiquitous homelinks based on IEEE 1394 and ultra wideband solutions. *IEEE Communications Magazine*, 41(4):74–82, April 2003.

105. 802.11 Standard. Draft supplement to standard for telecommunications and information exchange between systems – LAN/MAN specific requirements- part 11: Wireless MAC and PHY specifications: High speed physical layer in the 5 GHz band. *P802.11a/D6.0*, May 1999.

106. J. C. Harrtsen. The bluetooth radio system. *IEEE Personal Communications*, 7(1):28–36, February 2000.

107. K. J. Negus, A. P. Stephens, and J. Landsford. Homerf: Wireless networking for the connected home. *IEEE Personal Communications*, 7(1):20–27, February 2000.

108. R. J. Fontana, E. Richley, and J. Barney. Commericalization of an ultra wideband precision asset location system. In *UWBST 2003 IEEE Conference on Ultra Wideband Systems and Technologies*, November 2003.

109. E. M. Staderini. UWB radar in medicine. *IEEE Aerospace and Electronic Systems Magazine*, 17(1):13–18, January 2002.

110. T. E. McEwan. Body monitoring and imaging apparatus and method. *US Patent 5,766,208*, June 1998.

111. I. Y. Immoreev. Ultra-wideband (UWB) radar for remote measuring of main parameters of patient's vital activity. In *IEEE International Workshop on Ultra Wideband and Ultra Short Impulse Signals*, October 2002.

112. F. A. Duke. *Physical Properties of Tissue: A Comprehensive Reference Book.* Academic Press, 1990.

113. P. Charles and P. Elliot. *Handbood of Biological Effects of Electromagnetic Fields.* CRC Press, 1995.

114. I. Maravic and M. Vetterli. Sampling and reconstruction of signals of finite rate of innovation in the presence of noise. *IEEE Transactions on Signal Processing*, 53(8):2788–2805, 2005.

115. A. Taflove. *Advances in Computational Electrodynamics: The Finite-Difference Time-Domain Method.* Artech House Books, 1998.

116. O. P. Gandhi, B. Q. Gao, and J. Y. Chen. A frequency-dependent finite-difference time-domain formulation for general dispersive media. *IEEE Transactions on Microwave Theory and Technology*, 41(4):658–665, 1993.

117. IEEE Standards. http://www.ieee802.org/15/pub/.

118. Multiband Organization. http://www.uwbmultiband.org.

119. Multiband OFDM Coalition. http://www.multiband-ofdm.org.

120. IEEE802.15.4a. http://www.ieee802.org/15/pub/tg4a.html.

121. UWB Forum. http://www.uwbforum.org/.

122. Multiband OFDM Alliance (MBOA). http://www.multibandofdm.org/.

123. L. D. Nardis, P. Baldi, and M. D. Benedetto. UWB ad-hoc networks. In *IEEE Conference on Ultra Wideband Systems and Technologies*, May 2002.

124. F. Cuomo, A. Baiocchi, C. Martello, and F. Capriotti. Radio resource sharing for ad hoc networking with UWB. *IEEE Journal on Selected Areas in Communications*, 20(9):1722–1732, April 2000.

125. W. Horie and Sanada Y. Novel packet routing scheme based on location information for UWB ad-hoc network. In *IEEE Conference on Ultra Wideband Systems and Technologies*, November 2003.

126. C. Huitema. *Routing in the Internet*. Prentice Hall, 1999.

127. Y. B. Ko and N. H. Vaidya. Using location information in wireless ad hoc networks. In *IEEE Vehicular Technology Conference*, May 1999.

128. W. Horie, Y. Sanada, and M. Ghavami. Retransmission scheme with code sense for VSF/DS-UWB ad-hoc network. In *International Workshop on Wireless Ad Hoc Networks*, May 2005.

129. R. Ziemer, M. Wickert, and T. Williams. A comparison between UWB and DSSS for use in a multiple access secure wireless sensor network. In *IEEE Conference on Ultra Wideband Systems and Technologies*, November 2003.

130. S. M. Alamouti. A simple transmit diversity technique for wireless communications. *IEEE Journal on Selected Areas in Communications*, 16(8):1451–1458, October 2000.

131. V. Tarokh, N. Seshadri, and A. R. Calderbank. Space-time codes for high data rate wireless communications: Performance criterion and code construction. *IEEE Transactions on Information Theory*, 44(2):744–765, March 1998.

132. G. J. Foschini. Layered space-time architecture for wireless communication in a fading environment when using multiple antennas. *Bell Labs Technical Journal*, 1(2):41–59, 1996.

133. L. Yang and G. B. Giannakis. Space-time coding for impulse radio. In *IEEE Conference on UWB Systems*, May 2002.

134. F. Héliot, M. Ghavami, and R. Nakhai. Performance of space-time block coding and space-time trellis coding for impuse radio. In *IEEE Globcom*, November 2004.

135. F. Héliot, M. Ghavami, and R. Nakhai. Design and performance analysis of a space-time block coding scheme for single band UWB. *IEE Proceedings Communications*, 2006.

136. M. R. Hueda, G. Corral-Briones, and C. E. Rodriguez. MMSEC-rake receivers with resolution reduction of the diversity branches: Analysis, simulation, and applications. *IEEE Transactions on Communications*, pages 1073–1081, June 2001.

137. X. Chu and R. Murch. Performance analysis of DS-MA impulse radio communications incorporating channel-induced pulse overlap. *IEEE Transactions on Wireless Communications*, 5(4):948–959, April 2005.

138. A. Ghosh, D. R. Wolter, J. G. Andrews, and R. Chen. Broadband wireless access with WiMax/8O2.16: Current performance benchmarks and future potential. *IEEE Communications Magazine*, 43:129–136, February 2005.

139. J. Foerster and J. Liebetreu. FEC performance of concatenated Reed–Solomon and convolutional coding with interleaving. Technical report, IEEE 802.16 Broadband Wireless Access Working Group, June 2000.

140. R. Giuliano and F. Mazzenga. On the coexistence of power-controlled ultrawide-band systems with UMTS, GPS, DCS1800, and fixed wireless systems. *IEEE Transactions on Vehicular Technology*, 54:62–81, January 2005.

141. I. Gresham, A. Jenkins, R. Egri, C. Eswarappa, N. Kinayman, N. Jain, R. Anderson, F. Kolak, R. Wohlert, S. P. Bawell, J. Bennett, and J.-P. Lanteri. Ultra-wideband radar sensors for short-range vehicular applications. *IEEE Transactions on Microwave Theory and Techniques*, 52(9):2105–2122, September 2004.

Index

Printed and bound in the UK by
CPI Antony Rowe, Eastbourne

Printed and bound by CPI Group (UK) Ltd, Croydon, CR0 4YY

16/04/2025

14658562-0001